Multi-criteria Decision Analysis

Multi-criteria Decision Analysis

For Supporting the Selection of Engineering Materials in Product Design

Second Edition

Ali Jahan
Department of Industrial Engineering, Semnan Branch,
Islamic Azad University, Semnan, Iran

Kevin L. Edwards
Institution of Engineering Designers, Courtleigh, Westbury
Leigh, Wiltshire, UK

Marjan Bahraminasab
Department of Materials Science and Engineering, Buein
Zahra Technical University, Buein Zahra, Qazvin, Iran

ELSEVIER

AMSTERDAM • BOSTON • HEIDELBERG • LONDON
NEW YORK • OXFORD • PARIS • SAN DIEGO
SAN FRANCISCO • SINGAPORE • SYDNEY • TOKYO
Butterworth-Heinemann is an imprint of Elsevier

Butterworth-Heinemann is an imprint of Elsevier
The Boulevard, Langford Lane, Kidlington, Oxford OX5 1GB, UK
50 Hampshire Street, 5th Floor, Cambridge, MA 02139, USA

Notices
Knowledge and best practice in this field are constantly changing. As new research
and experience broaden our understanding, changes in research methods, professional
practices, or medical treatment may become necessary.

Practitioners and researchers must always rely on their own experience and knowledge in
evaluating and using any information, methods, compounds, or experiments described herein.
In using such information or methods they should be mindful of their own safety and the
safety of others, including parties for whom they have a professional responsibility.

To the fullest extent of the law, neither the Publisher nor the authors, contributors,
or editors, assume any liability for any injury and/or damage to persons or property
as a matter of products liability, negligence or otherwise, or from any use or operation
of any methods, products, instructions, or ideas contained in the material herein.

ISBN: 978-0-08-100536-1

British Library Cataloguing-in-Publication Data
A catalogue record for this book is available from the British Library.

Library of Congress Cataloging-in-Publication Data
A catalog record for this book is available from the Library of Congress.

For Information on all Butterworth-Heinemann publications
visit our website at http://store.elsevier.com/

Working together
to grow libraries in
developing countries

www.elsevier.com • www.bookaid.org

Contents

Preface to the 2nd edition

Since publication of the first edition of this book, there has been a considerable increase in interest in the subject of multi-criteria decision analysis applied to materials selection and design. Originally, it was thought that the comprehensive collection of Multi-Criteria Decision Making (MCDM) methods described in the book would be sufficient to remain complete and up-to-date for many years. However, the rate of change has been more rapid than expected, with some new and important areas of research activity occurring recently that need to be taken into account to more accurately reflect the state of the art. This was the motivation to produce a new edition of the book with much greater scope and updated content. To support this work, it was necessary to include a third author, Dr Bahraminasab, with expertise in material design and analysis.

Effective decision-making is the basis of successfully solving any engineering design problem, including selecting the most appropriate materials to use. Rapid changes in manufacturing technologies also provide opportunities for developing new materials. To meet the increasing expectations of customers for more sophisticated products, multi-functional hybrid materials can also be designed. MCDM can support the process of selecting appropriate materials from long a list of options or in selecting the constituents for hybrid materials. MCDM methods can be divided into Multi-attribute Decision Making (MADM) methods and Multi-objective Decision Making (MODM) methods. MADM methods are used for materials selection problems as well as design selection and process selection problems. MODM methods can be used for material design and optimization problems as well as simultaneous material and design optimization studies.

This new edition builds on the success of what was the first e-book published on the subject three years ago, and describes the latest research, including the authors' own work related to the subject. Although primarily aimed at academics, for a better understanding by practitioners, this new edition describes the benefits of MCDM for various applications including materials selection, material design, and design selection to demonstrate the strategic application of the methods through a greater number of more detailed and diverse case studies.

The key changes in this new edition are the addition of two completely new chapters, revised and expanded existing chapters with additional new sections, and more detailed case studies. To enhance the structure of each chapter and guide the reader better, Abstract, Keywords, Learning Aims, and Summary and Conclusions sections have been added. More review questions have also been included to ensure the learning aims for each chapter can be fully assessed from the content.

The new chapter "Materials selection in the context of design problem solving" provides an overview of systematic and creative design tools and principles in material design and materials selection. Particular attention is given to the use of Quality Function Deployment, "Lean" design principles, and creative problem-solving. The new chapter "Multi-objective decision-making for material and geometry design" describes materials tailoring as well as material and design optimization. There are also new sections in several chapters on "Combining functional and human centered attributes in materials selection," "Cost considerations of materials selection/substitution," "Use of new hybrid materials as opposed to currently available materials," "Graphical based ranking methods in material selection," "Principal attributes of normalization," "Classifying dimensionless methods," "An introduction to fuzzy MADM methods for material selection," and "Final remarks on trends and challenges for the future." The number of case studies has also increased considerably from three to ten detailed practical mechanical engineering examples and represents the largest chapter in the book.

The methods described in this book again draw on the published work of Prof. Ashby as well as other prominent researchers in the fields of materials selection, multi-criteria decision analysis and design, including (alphabetically) the published work of Profs. Abedian, Brechet, Chakraborty, Chatterjee, Farag, Mahmudi, McDowell, Milani, Rao, Sapuan, Shanian, Silva, and many others cited in the extensive and updated reference lists found at the end of each chapter. The reference lists allow researchers to easily locate further information on specific topics of interest.

This extensive rework of the book has benefitted this time from the support of Islamic Azad University, Semnan Branch, through an internal grant, and the University is thanked for the close cooperation. In conclusion, the authors would again like to acknowledge colleagues for their helpful comments and express thanks to their families for the support and encouragement given during the preparation of this new edition.

Ali Jahan, Kevin L. Edwards and Marjan Bahraminasab
December 2015

From the Preface to the 1st edition

The aim of this book is to provide the means for designers to produce successful engineering product design by improving decision-making when choosing materials. The choice of the best material from a range of alternative materials greatly impacts the eventual technical success of a product and avoids undesirable cost implications and possible premature failure in the market place. An improper selection of material can adversely affect productivity and profitability, and even undermine the name of an enterprise because of the growing demands for through-life and extended manufacturer responsibility.

It is a constant challenge for designers, even when educated in the fundamentals of materials and mechanical engineering, to be able to optimally select materials to satisfy complex design problems, when faced with the vast range of materials available. Current approaches to materials selection range from the use of intuition and experience to more formalized methods. The latter is also available in computer-based form and includes electronic databases with search engines to facilitate the materials selection process. Recently, multi-criteria decision-making (MCDM) methods have begun being applied to materials selection, demonstrating significant capability for tackling complex design problems. Therefore, this book describes the growing field of MCDM applied to materials selection and has been prepared to assist academic teaching and research as well as practitioners involved in materials engineering and design.

The methods described in the book draw on the published efforts of other prominent researchers in the field, particularly the work of Prof. Mike Ashby. The current approaches developed and new ideas proposed complement and enhance existing materials selection and MCDM methods and when combined strategically help to ensure a more informed choice of materials if the design problem being analyzed is complex in nature.

Lecturers in materials engineering need to be able to explain to students how suitable materials can be appropriately selected to satisfy a given set of engineering design requirements. Currently a lot of universities and colleges use materials selection charts (the "Ashby" approach) and its computer software implementation, the CES Selector from Granta Design Ltd, to help teach materials selection. The use of MCDM supplements the capability of the materials selection chart method, allowing more advanced design applications to be considered. There are a lot of printed materials books available, but most are presented from a materials science perspective. There are fewer books available on materials engineering, with some containing a chapter on materials selection. There are even less books devoted solely to materials selection. Not many of these books address the subject quantitatively, limiting their application to simple component design. To the best of the authors' knowledge there are currently no e-books published on materials selection and this book aims to supplement the materials selection chart approach.

The book is designed for researchers and decision makers in applying MCDM methods, especially multi-attribute decision making methods, and will be of interest to students, academics, and designers in the different fields of engineering and design. This book will be useful for final stage undergraduate and postgraduate level teaching of materials, mechanical engineering and product design programs. It will be invaluable for researchers in materials engineering disciplines and in operational research, specifically multi-criteria decision analysis. It can also be used as a reference text for engineering practitioners to help select the best material(s) for advanced design applications.

Chapter "The importance of decision support in materials selection" explains the position of materials selection in the engineering design process. Chapter "Materials selection in the context of design problem-solving" gives an overview of screening methods used in materials selection. Chapter "Screening of materials" provides an introduction to the use of MCDM for materials selection. Chapter "Multi-criteria decision-making for materials selection" describes methodologies for supporting enhanced decision-making in materials selection and supplements existing quantitative methods by allowing simultaneous consideration of design attributes, component configurations, and types of material. Chapter "Multi-attribute decision-making for ranking of candidate materials" provides practical case studies in biomedical and aerospace engineering applications, demonstrating the scope and use of the methods. The above mentioned chapters conclude with a set of review questions to help readers check their understanding of the subject. The last chapter concludes with suggestions for further research to support advanced decision making in materials selection, allowing the full potential of MCDM to be realized.

The methods described in the book have been implemented in prototype form as spreadsheet and mathematical analysis tools but have the potential to be implemented as commercial computer software; either stand-alone linked to electronic materials databases or integrated into existing search engines used in materials selection. Some of the work described is in a developmental state and therefore notification of possible errors and areas of further improvement are most welcome. It is hoped that readers will find the ideas presented to be interesting and helpful when put into practice or in further studies.

Finally, the authors would like to acknowledge colleagues for helpful comments and thank wives and families for the support and encouragement given during the preparation of the book.

Ali Jahan and Kevin L. Edwards
December 2012

About the authors

Ali Jahan

Dr Jahan is active in the development of multi-criteria decision-making (MCDM) techniques for the engineering design process, especially for materials selection. His main research interests are in the application of MCDM and quality tools for improving materials, design, and processing. Dr Jahan has been involved in different projects in industry related to quality engineering and designing systems based on the requirements of quality management standards and associated decision support systems. He has various publications describing the development of MCDM in design and materials engineering. Dr Jahan regularly reviews papers in different international journals, and is the recipient of awards related to his research and academic activities. He is Assistant Professor in the Faculty of Industrial and Mechanical Engineering, Semnan Branch, Islamic Azad University, Iran.

Kevin L. Edwards

Professor Edwards leads research and development, and knowledge transfer activities in materials engineering and design technology innovation, with a particular emphasis on the relationship between materials, processing, and design. He also advises government, businesses, and universities on engineering educational and professional standards development, as well as being an active member of several professional engineering institutions. Professor Edwards has had a diverse career with senior appointments in higher education, aerospace and automotive industry, and engineering consultancy. His broad experience of technical and academic management, business and curriculum development, and quality assurance underpins teaching, research, and professional practice in advanced composite materials and processes, engineering product design and analysis, and innovation management, with an expertise in design decision support, systematic materials selection, and new product development. Professor Edwards has been awarded high value research grants and industrial funding for numerous large-scale collaborative multi-disciplinary research programs, resulting in the development of novel materials, engineering products, manufacturing systems, and computer-based design tools. He publishes regularly in periodicals, conference proceedings, and journals, has participated in international conference organizing and scientific committees, chaired seminars and workshops, given keynote presentations, and is a member of the editorial boards of several journals.

Marjan Bahraminasab

Dr Bahraminasab is Assistant Professor in the Department of Materials Science and Engineering at Buein Zahra Technical University, Qazvin, Iran. She received a PhD in biomaterials engineering from University Putra Malaysia, and is active in the application of multi-criteria decision-making in materials engineering and design. Dr Bahraminasab's research on the design of a new biomaterial for knee prostheses was featured in Asia Research News. Her areas of interest include biomaterials and biomechanics, materials design and selection, and finite element analysis.

The importance of decision support in materials selection

1

Learning Aims

The overall aim of this chapter is to gain an overview of decision-making involved in materials selection. After carefully studying this chapter you should be able to understand:

- the critical stages of the materials selection process associated with designing new products
- the significance of supporting effective decision-making in materials selection
- the notion of relationship between design, materials, and manufacturing processes
- the challenges of selecting materials to satisfy conflicting design requirements
- the economic considerations that should be taken into account when selecting materials
- the approaches that can be used for substituting either conventional or new materials in product design
- the role of materials in the creation of sustainable products.

1.1   Introduction to materials selection

The selection of the most appropriate material for a particular purpose is a crucial function in the design and development of products. Materials influence product function, customer satisfaction, production systems, product life cycle, who is going to use or produce it, usability, product personality, operating environment, and costs in a complex way.

Materials selection can be carried out to either choose alternative materials for changes to the design of an existing product in order to reduce say cost or weight, meet new legal requirements, overcome failure occurrence, or satisfy different market demands, or it can be used to choose materials for the design of a completely new product. The materials selection process is similar for existing and new products although the starting point and information requirements may differ. The interdisciplinary effort required in most cases is nontrivial and the engineering designer not only requires detailed, accessible, and timely information about materials' properties but also knowledge of multi-criteria decision-making (MCDM).

This book describes the main principles and strategic application of MCDM techniques to support engineering product designers compare the performance of established materials, hybrid materials, and new materials, when selecting the most appropriate materials for new product design.

Multi-criteria Decision Analysis. DOI: http://dx.doi.org/10.1016/B978-0-08-100536-1.00001-1

1.2 Background and justification for formalized materials selection

There are an enormous number of materials available, each with a range of different properties and behaviors. New materials are also constantly being developed with enhanced properties, expanding the list of options available to the engineering designer. The materials' properties, and combination of properties in the form of performance indices, can be mapped on to materials' selection charts, pioneered by Ashby (2013). In the charts, the materials naturally cluster into the different classes of metals, polymers, elastomers, glasses, and ceramics. However, only parts of the charts are populated with materials, leaving holes or gaps in the selection space. New materials with enhanced properties can reduce the size of or fill gaps within clusters, or expand the boundary of clusters, or the gaps between clusters can potentially be filled or "bridged" with hybrid or multi-materials such as composite materials as shown schematically in Fig. 1.1.

The historical evolution in the use and development of materials reflects the progress of the interdisciplinary science from the early civilizations until today (Brechet and Embury, 2013). The strategy with respect to materials usage started in the Stone Age by using the available materials such as stone and wood and then later copper and bronze (Bronze Age) and iron (Iron Age). Afterwards, the strategy gradually focused on the optimization of specific classes of materials. This lead to the development of tools for comparing and selecting materials from different classes of materials already optimized in terms of their engineering potential. Today, the emphasis has shifted more towards the consideration of economical aspects and environmental impact. This has created a tendency towards the development of materials using design strategies with an increased importance of modeling and multi-functionality of materials (Brechet and Embury, 2013). However, there is still

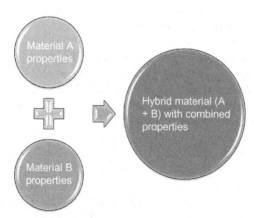

Figure 1.1 Hybrid materials combine the properties of two (or more) monolithic materials.

a lot of fundamental materials research being conducted without careful consideration being given to its practical application (Edwards, 2011). This not only justifies the need for the greater use of materials selection tools but also the importance of supporting decision-making to better understand and manage the multi-objective product design process.

1.3 Decision-making and concession in product design

Introducing a completely new product or improving an existing product involves a complex chain of interdependent activities including design, analysis, materials selection, and consideration of manufacturing processes, and all depend on MCDM. As well as influencing material properties, process selection is a prerequisite to manufacturing equipment selection. However, materials selection used to be only a minor part of the design process (Chiner, 1988) and therefore has not received the same level of research and development as other fields of design. The selection of suitable materials for a specific purpose though is a difficult, time-consuming, and expensive process because of the large number of available materials with complex relationships and various selection parameters. As a consequence, approximations are made with materials frequently being chosen by "trial-and-error" or simply on the basis of what has been successfully used in the past. This approach though can lead to compromise and unpredictable outcomes, possible premature failures, and limits the ability to achieve an optimal choice of materials.

The stage reached in the design process is important because the nearer a product is to manufacture the greater is the cost of making any design change (Charles et al., 1997). It has been estimated that the relative cost of a design change after manufacture is 10,000 times more than at the conceptual stage of design (Charles et al., 1997). Therefore, it is worth making decisions carefully and spending enough time early on in the design process using systematic selection techniques. This makes it easier to manage the "trade-offs" between design, materials, shape, and manufacturing processes and lead to an optimum design solution.

To satisfy customer requirements, manufacturing organizations must be continually aware of product costs, reliability, durability and recyclability, and market trends. These attributes should be addressed strategically by manufacturers through a continuous process of improvement in an ongoing effort to improve their products (Jahan and Edwards, 2013). This will only be fully achieved through optimum decision-making about design, materials, and manufacturing processes (Chakladar et al., 2009; Blanch et al., 2011) and provides the opportunity for sustainable and profitable growth. To support this process, MCDM techniques have developed dramatically in both theory and practice, especially in the fields of design and manufacturing, with growing interest in their application to materials selection.

1.4 The position of materials selection in the engineering design process—from concept to detail stages

The successful design of an engineering component is integral to satisfying the functional and customer specified requirements for the overall product of which it forms a part, utilizing material properties and capabilities of suitable manufacturing processes (Zarandi et al., 2011). The behavior of a material used to create a component will be affected by component geometry, external forces, properties of stock material before processing, and the effect of manufacturing (or fabrication) method (Farag, 2008a). The evaluation of the typically large number of design solutions (altering the size, shape, and mass of the component) and suitability of an even larger number of different materials rapidly becomes too complicated to be intuitive. This highlights the value of being able to use MCDM to support decision-making in the engineering design process. Although experimental-based selection of a material, eg, testing, prototyping, etc., for a specific design solution is the most accurate, it quickly becomes unreasonable due to the time required and the high costs of experiments, especially if several materials have to be considered. Alternatively, other more viable options can be considered such as computer-based simulations but the ranking of materials should have already been successfully completed during the initial stage of the design process (Jahan and Edwards, 2013). Fig. 1.2 shows connection of materials selection and continuous improvement in product development. There is the option to develop a new material but this adds cost and risk to the design efforts and has been highlighted here for the occasion when there is no acceptable material options (Michael, 2009; Edwards, 2008). However, the added cost and risk may be worthwhile if there is an innovation that provides a product with a competitive advantage compared to products produced by other companies. In Fig. 1.3, quality tools such as quality function deployment (QFD) (Mayyas and Shen, 2011; Kazemzadeh et al., 2009) helps marketing and design teams incorporate the voice of the customer in product designs, increasing

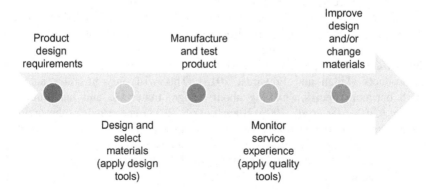

Figure 1.2 Materials selection and continuous improvement in product development.

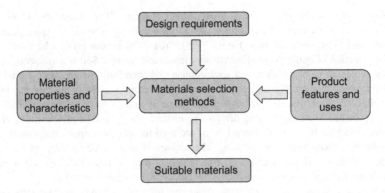

Figure 1.3 Combining materials and product information in the materials selection process.

the likelihood that the final product will successfully satisfy the customer's needs. Failure modes and effects analysis (FMEA) (Chin et al., 2008) will indicate the material's attributes that must be investigated, controlled, and monitored to ensure the reliability of the component in which the material is used. It will also indicate the manufacturing process steps and controls that are required to make a product, subassembly, or component that will consistently meet its design requirements. Therefore, QFD and FMEA, when used as quality tools in design, can dramatically improve product development efficiency because they lead to systematic design, materials selection, and manufacture.

Ashby (2005) and Ashby et al. (2013) map the application of design tools and materials selection on the traditional systematic product development process. The resulting methodology is a "design-led" approach to materials selection. Computer-based design tools, of which there are many, enable visualization (eg, 2D Computer-Aided Design (CAD)), modeling (eg, 3D solid modeling), and analysis (eg, finite element analysis (FEA)) to be conducted, as well as interactive simulation of evolving design solutions. To ensure that product design, materials, and process selection are done together, Design for Manufacture and Rapid Prototyping tools are also regularly used. There are fewer tools available that are dedicated to materials selection, and even less for materials process selection, although many design tools make use of materials properties databases. At each stage, from concept to detail design, information about the materials gradually changes in the level of detail and precision, as the number of available materials reduces from initially all possible materials to finally one material.

Fundamentally, the strategy for selecting materials is determined by the need to satisfy design requirements and involves five main stages:

1. Definition of the design (Chiner, 1988).
2. Analyzing and translating the design requirements (expressed as constraints and objectives) into material performance indices/properties (Ashby et al., 2004; Dieter, 1983, Chiner, 1988; van Kesteren et al., 2006).
3. Comparing the properties required with a materials property database in order to select a few candidate materials that look promising for the application (van Kesteren et al., 2006,

Jalham, 2006; Ashby et al., 2004; Dieter, 1983, Chiner, 1988; Giaccobi et al., 2009; Farag, 2002). Screening is based on the idea of classifying performance requirements into "rigid" and "soft" requirements (Farag, 1979). "Rigid" requirements can be used for the initial selection of materials to eliminate any unsuitable groups. Soft requirements are normally subjected to negotiation and compromise and therefore involve more complicated decision-making. Examples of requirements in this group consist of mechanical properties, physical properties, and cost, which are normally compared on the basis of their relative importance to an engineering design. For instance, high-specific stiffness and strength materials such as fiber/polymer composites are used for advanced sports equipment.

4. Developing, comparing, and ranking alternatives (Farag, 2002; Ashby et al., 2004; Jalham, 2006; van Kesteren et al., 2006) and using "soft" material requirements to further narrow the field of possible candidates down to a few promising materials.

5. Searching for supporting information about the top-ranked candidate materials (Ashby et al., 2004), evaluation and decision for the optimal solution (Dieter, 1983; Chiner, 1988, 2006; Farag, 2002), and finally verification tests (Chiner, 1988).

1.5 Understanding the functional requirements and design criteria in selecting materials

The performance requirements of materials can be classified into two main categories: "rigid," or "go-no-go" requirements and "soft," or relative requirements (Farag, 1979). "Rigid" requirements should be met by the material if it is to be considered at all. Such requirements can be used for the initial selection of materials to eliminate unsuitable classes of materials. For instance, metallic materials are eliminated when selecting materials for an electrical insulator. "Soft," or relative requirements, are subject to negotiation and trade-offs.

Apart from materials' properties, other requirements, such as those on materials processing during component manufacturing and those on in-service conditions relating to the materials of components, should also be taken into account in materials selection (Edwards and Deng, 2007).

The design of a product is either cost driven or performance driven and the approach taken makes a significant difference when choosing materials. An example of a cost-driven product is a mass produced motorcar or beverage can/bottle and an example of a performance-driven product is a bespoke biomedical prosthesis/implant or a tennis racquet. The selection of cheaper materials may lead to reduced manufacturing organization cost but the selection of lighter materials although more expensive, in say a motorcar or aircraft, may reduce fuel consumption, service cost, and environmental pollution.

1.6 Combining functional and human-centered attributes in materials selection

The process of selecting materials for the design of products is complicated because of the need to match a set of contradictory design requirements, both functional and human-centered (user) attributes, to a range of material properties and characteristics (sensorial qualities), all having economic implications (Brechet et al., 2001). This often leads to compromise in the design and choice of materials, resulting in poor products and user satisfaction (Ljungberg and Edwards, 2003). There are methods available to help select materials and these have proved effectiveness for product functional requirements. The methods are mostly systematic and numerically based, with some implemented as computer software packages, and depend on matching the desired material property attributes against those from available materials. It is also relatively easy to access a lot of detailed technical information that is available from many different sources. There is much less support available for human-centered requirements despite it being equally as important to the success of a product. This is mainly because human-centered requirements are not conducive to formal systematic procedures, and information on material characteristics is much less available. Also, the properties and characteristics of materials are interrelated and ideally should be considered together (Ulrich and Eppinger, 2004).

Most products are differentiated by their technical functions and by what the materials they are made of mean to the user (Karana, 2006). However, unlike functional requirements, which are defined in quantitative terms and can be assessed objectively, human-centered requirements are expressed in qualitative terms and hence are more subjective and difficult to interpret (Schifferstein and Hekkert, 2007). Also, the lack of structured approaches can lead to unpredictable outcomes, with material characteristics not being given sufficient priority in the design process. There is therefore a lot of reliance on the skill and experience of product designers when selecting materials. For effective materials selection in product design, both functional and human-centered requirements need to be thoroughly considered (Karana et al., 2008). Unfortunately, the properties and characteristics of materials are often divergent, which regularly results in conflict when trying to satisfy all the design requirements. Therefore, priorities have to be given to the design requirements that are considered to be essential and restricted by what materials will allow.

The materials selection process essentially converts a set of design requirements into a list of suitable materials as shown in Fig. 1.3. To be successful, the process relies heavily on access to information on different material properties and features and product features and uses. However, unlike material properties, material characteristics are much more difficult to quantify, and hence more challenging to locate, sort, and use appropriately. There are similar problems with the use of existing product design features, or use of materials in products, because they are difficult to categorize for retrieval purposes and therefore mostly rely on intuition and experience when comparing with new product designs. This regularly leads to new products being designed within the confines of the existing knowledge of materials and products only, and as a consequence stifling design innovation.

The materials selection methods used for functional requirements are based on systematic analysis. The methods depend on precise information, models, and sets of rules that can be manipulated to accurately match design requirements with material properties. The materials selection methods used for human-centered requirements are based on analogy and synthesis. The methods rely on past experience and the ability to match desired features with those of previous design solutions. However, for a product to be successful, the different methods of materials selection have to be combined strategically and used appropriately to allow decisions to be made that are sensitive to the changing nature of the selection problem as the design evolves (van Kesteren, 2008). However, if the information is quantifiable, analysis can be used and if the information is qualitative, synthesis can be used. The materials selection process will therefore alternate between analysis- and synthesis-based methods and this needs to be carefully managed to help ensure consistency in decision-making. The MCDM technique can be used to provide overall support of the materials selection process by allowing the simultaneous consideration of design requirements, material properties and characteristics, and product features and uses (Jahan et al., 2010).

1.7 Cost considerations of materials selection/ substitution

When considering the factors that decide the success of a product in the market today it becomes clear that cost is as crucial as quality and functionality (Layer et al., 2002). The term cost means the sum of money expended in terms of labor, materials, use of equipment, etc., to produce a product (Layer et al., 2002). Product design, especially the selection of materials, can highly contribute to total product cost (Asiedu and Gu, 1998). The choice of materials has implications throughout the life cycle of a product, influencing many aspects of economic and environmental performance. It has impacts on the consumption of raw materials and energy (Thurston and Locascio, 1994). Economic advantage can be achieved as a result of introducing cheaper materials, more cost-effective use of materials, lower cost of processing, better recyclability and lower cost of disposal, or lower running cost of the product (Farag, 2008b). Although both design and economic justifications have the common goal of arriving at a competitive product, their goals are diametrically opposite to each other (Noble and Tanchoco, 1990). Therefore, cost can be employed as an evaluation criterion in design either in a design-to-cost or design-for-cost context. Design-for-cost is the conscious use of engineering process technology to reduce life cycle cost (LCC), while design-to-cost creates a design satisfying the functional requirements for a given cost target. LCC analysis provides a framework for specifying the estimated total cost. The growing demand to develop products that are inexpensive has necessitated the consideration of product LCC during the design stage. LCC includes all the costs incurred through product creation, use, and disposal. However, during the materials selection stage, the use of LCC analysis should be restricted to the cost that can be controlled (Asiedu and Gu, 1998). Estimating the LCC of a proposed product during its development phase is required for a number of reasons (Asiedu and Gu, 1998):

1. Determining the most cost-efficient design amongst a set of alternatives.
2. Determining the cost of a design for budgetary purposes.
3. Identifying cost drivers for design changes and optimization.

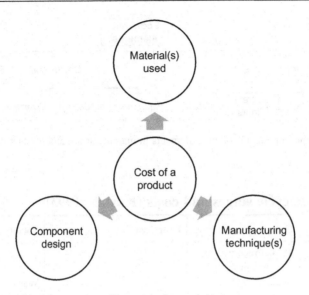

Figure 1.4 Interrelated factors that affect cost of a product.

Fig. 1.4 shows three interrelated factors that affect the cost of a product. Component design is the specification of size, shape, and configuration, and some part of the cost of a component is associated with its design.

In terms of economics, it is necessary to select the material(s) with the required properties that are the least expensive. Material price is usually quoted per unit mass. The component volume may be determined from its dimensions and geometry, which is then converted into mass using the density of the material. In addition, during manufacturing there is some unavoidable material waste, which should also be taken into account in the calculations. As already stated, the choice of manufacturing process will be influenced by both the material selected and component design. The major cost considerations for manufacturing include capital equipment, tooling, labor, repairs, machine downtime, and waste as shown in Fig. 1.5 (Callister and Rethwisch, 2007). The rate of production is also a significant consideration, as well as costs associated with inspection and packaging of the final product. There are other factors, including labor fringe benefits, supervisory and management labor, research and development, property and rent, insurance, profit, and taxes, that are not directly related to design, material, or manufacturing but also influence the selling price of the product. In the context of materials substitution for high volume production, the problem of how to accurately predict the costs are of the highest importance in today's highly competitive industries (Henriques et al., 2014).

The LCC of a product is made up of the costs to the manufacturer, user, and society as shown in Table 1.1. While the LCC is the aggregate of all the costs incurred in a product's life, there are differences between the cost issues that will be of interest to the designer of the product and the company developing the product in an LCC

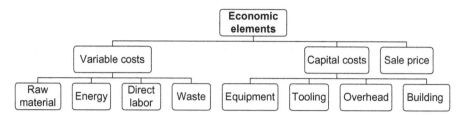

Figure 1.5 Some important economic elements in manufacturing and product development.

Table 1.1 Life cycle stages and costs (Kusiak, 1993)

	Company cost	User cost	Society cost
Design	Market recognition, Development		
Production	Materials, consumption of energy, facilities, wages		Waste, pollution, health damages, emissions to the environment
Usage	Transportation, storage, waste, breakage, warranty services	Transportation, storage, materials, energy efficiency of final product, maintenance	Packaging, waste, pollution, health damages
Disposal/ Recycling		Disposal/recycling fees	Waste, disposal, pollution, health damages

analysis. While the company must know the total cost of the product, the designer is only interested in the costs that can be controlled.

With regard to the time of use, three different types of calculation are distinguished for cost factors (Normung, 1989; Layer et al., 2002):

1. Precalculation.
2. Intermediate calculation.
3. Postcalculation.

Precalculation is only able to access product describing data, and unfortunately such data may be incomplete or uncertain. The use of unreliable data necessitates suitable decision-making methods for design selection. During the product development cycle, intermediate calculations are carried out and then applied to control costs. Therefore, because the methods for precalculation come mainly from the field of engineering sciences, intermediate and postcalculation methods have come from business administration concerns (Layer et al., 2002). Cost decision certainty seldom exists in a competitive society (Asiedu and Gu, 1998), and this is one issue

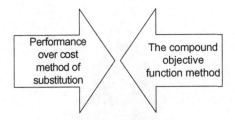

Figure 1.6 Methods for economic material selection/substitution.

that has been largely ignored in cost estimating models. Different methodological approaches for cost estimation during the product development process are described and classified in a scientific context in the literature. Depending on the approach chosen, cost estimation methods provide either a qualitative or a quantitative result (Layer et al., 2002). Although raw material costs are normally easily available, second-order factors such as tools and dies cost, amount of scrap, rework and others, are quite difficult to predict to support the substitution decision (Henriques et al., 2014). An overestimated cost will prevent a company from remaining competitive, while an underestimated cost will result in losses.

Most products comprise of an assembly of several individual components, each of which have a cost. Fig. 1.6 shows two commonly used methods for economic material selection/substitution for a component (Farag, 2008b). The first method separately compares the performance (P) and total cost (C_t) of a candidate material against a currently used material. Parameter P covers all requirements of the material except for cost, and it is estimated using weighted properties method (Farag, 2002). As presented in Eq. (1.1), C_1 is the cost of material used in making the component; C_2 is the cost of manufacturing and finishing the component; C_3 is the running cost over the entire life of the component; and C_4 is the cost of disposal and recycling.

$$C_t = C_1 + C_2 + C_3 + C_4 \tag{1.1}$$

If the purpose of substitution is to improve performance and to reduce the total cost of a component, the one that gives the higher (P/C_t) over the currently used material is selected. The compound objective function method (Farag, 2008b) compares the compound objective function of the candidate materials against the currently used material. The performance index is defined as the weighted sum of all the normalized material performance requirements, including total cost (C_t).

1.8 The relationship between materials selection and processing

The selection of a material must be closely coupled with the selection of a manufacturing process. Manufacturers are always on the lookout for new materials and improved processes to produce better products, and thus maintain their competitive edge and

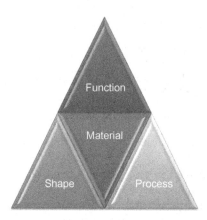

Figure 1.7 The interaction between material, function, shape, and process.

increase their profit margin (Farag, 2008b). Recent developments in manufacturing techniques are facilitating the development of new materials and improving the properties of current materials for different engineering applications. Consequently, the number of materials and new manufacturing processes is constantly on the rise with two main objectives: low cost and high performance (Jahan, 2012). Any change in a material will likely cause a change in the manufacturing process, geometry, shape, and performance (Fig. 1.7). Often there are many ways of creating components (shaping, joining, finishing, etc.) but choosing the optimum route is difficult (Edwards and Deng, 2007). With the increasing level of sophistication in both materials and processing, it is insufficient to just rely on experience alone (Lovatt and Shercliff, 1998). The large number of materials and the growing number of processes and subprocesses available to engineering designers, coupled with the complex relationships between different selection parameters, often make the selection of materials for a given component a difficult task especially when there are cost limitations.

Ashby (2005) extended his concepts of materials selection charts for properties at the conceptual design stage to the development of process selection charts. A lot of attention has been directed in recent years towards developing design methodologies for manufacturing and assembly for enhancing manufacturability and integrating material and process concepts through proper design with the aim of reducing the cost, enhancing the product quality, and enhancing the speed of delivery.

The production of tailored materials requires the development of new processing routes or important changes to conventional routes (Brechet and Embury, 2013). This subject is strongly important with regards to the new tendency for multifunctional materials, modeling, and computational materials engineering. The goal is to select a material and process that maximizes the quality and minimizes the cost of product. The design and manufacturing process, except for the simplest of products, is very complicated and involves making thousands of decisions in which any mistake can lead to unnecessary activities, extra cost, and customer dissatisfaction. Therefore, achieving optimum decision-making in the integrated chain of actions for developing new, or improving existing, products is essential (Jahan and

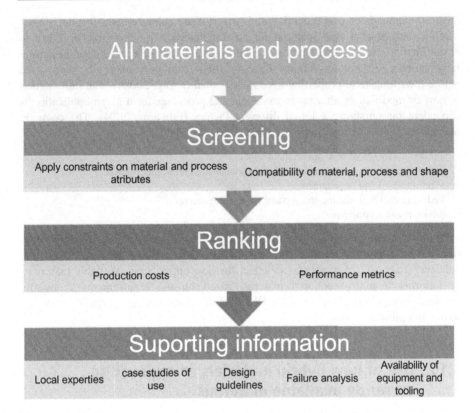

Figure 1.8 The different stages for materials and process selection in design.

Edwards, 2013). Fig. 1.8 shows the stages of material selection and its integration with process selection (Ashby et al., 2004).

Recently, the following sequence of tools was proposed to facilitate selection processes as an integrated material and process selection framework (Albiñana and Vila, 2012):

1. Documentation management that allows ideas to be translated into engineering design.
2. Matrices that relate incompatibility between materials and manufacturing processes.
3. Matrices that relate typical properties to specific requirements.
4. Models of manufacturing processes that take into account limitations in the geometry of components.
5. Multi-criteria analysis techniques.
6. Screening preselection techniques.

1.9 The significance of design adaptation and materials substitution

There are many traditional materials that have served in engineering applications for a long time but are now being replaced by so-called "new materials" in order to meet the demands of cost reduction and performance enhancement. If a decision is taken to substitute a new material for an established one, care must be taken to ensure that all the characteristics of the new material are well understood because a large number of product failures have resulted from new materials being used before their long-term properties were fully known (Farag, 2008b). The introduction of new or modified or alternative materials and processes for a given application is dependent on satisfying a lot of different factors (Edwards, 2004). The common reasons for materials substitution include (Edwards, 2004):

- Taking advantage of new materials or processes.
- Improving service performance, including longer life and higher reliability.
- Meeting new legal requirements.
- Reducing cost and making the product more competitive.
- Environmental constraints.

With the large range of properties that describe any material, it would be very rare to find a material that has the absolute ideal values for a function. Therefore, either a "trade-off" of properties to select the best option from available materials, or tailoring a multi-functional material based on the requirements is usually required. Both approaches need MCDM to deal with conflicting objectives design to obtain optimum performance.

1.10 Use of new hybrid materials as opposed to currently available materials

Various materials have been developed to meet the demand for enhanced functioning and providing improved quality of life with reduced cost. The rapid growth in the development of new materials is evidence of the vulnerability of current materials in fulfilling the desired design requirements. Material engineers have developed hybrid materials that are comprised of two or more different materials, designed in such a way as to offer attributes not provided by any one material alone (Ashby, 2013). New materials are being created based on two distinct strategies (Fig. 1.9):

1. The conventional strategy in which the goal is to provide a homogeneous and uniform structure, and most available materials are developed based on this strategy.
2. The recent novel strategy that aims at distributing the constituents in a selectively heterogeneous and controlled way to provide multi-functions and address conflicting requirements in niche applications (Brechet and Embury, 2013).

In the first strategy, homogenization involves a compromise between the desirable characteristics of the material required for a specific application. Examples of the materials developed by this strategy are engineering alloys and polymers in which the new materials are generated by managing the grain size, precipitation, crystallization state, and design of polymer chain and interchain bonding, respectively. These materials usually possess some improved properties at the expense of a reduction in other characteristics. The heterogeneous materials designed based on

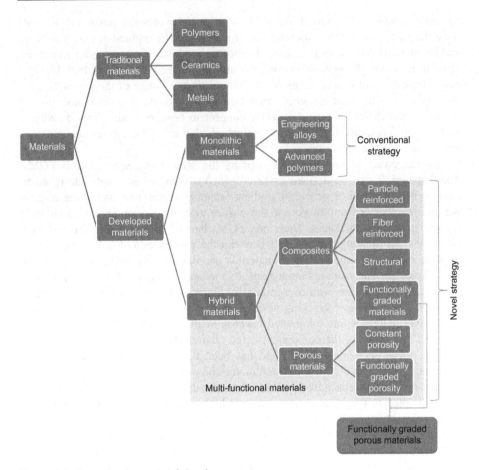

Figure 1.9 Strategies for material development.

the second strategy includes fiber and particle reinforced and structural composites (combination of two materials), porous materials (cellular materials), and the most recent, functionally graded materials or FGMs (combination of two or more materials, of which one could be cellular).

Composite materials provide some superior characteristics, including high specific modulus (modulus over density) and high specific strength (strength over density), which makes them useful materials in a large number of industrial applications. In the design of composite materials, the main concerns are well bonded and durable interfaces between the matrix and reinforcement. For example, in dentistry, different fiber-reinforced epoxy matrix composites have recently been used where the fibers have been made of carbon, polyamide, ultra-high molecular weight polyethylene, or glass, individually or in combination. Among these, glass fiber—reinforced polymer has shown properties (strength and modulus) comparable to that of dental tissues, in addition to good adhesion to the polymer matrix and aesthetic appearance (Khan et al., 2015). Porosity provides the possibility to optimize the stiffness, strength, and density of a material. In applications such as bone and tissue scaffolds, porosity reduces the elastic modulus of the material and at the same time allows the penetration of cells into the material (matching both biological and mechanical properties to that of natural tissue). Some of the presently used materials, such as stainless steels, titanium alloys and alumina ceramics, are not bioactive and therefore cannot strongly connect to bone or tissue. This drawback, therefore, limits their use in clinical applications due to the nonactive bond with tissue in the human body.

Another group of materials, designed by the second strategy above, is FGMs. These materials can be defined as heterogeneous composites usually designed to have a compositional or structural gradient selectively from one side of the component to the other or from the core to the surface, resulting in continuous variation in properties through thickness. Therefore, FGMs benefit from the pure form of each constituent eliminating the trade-off between the properties and utilize the characteristics of all constituents. These materials are finding a broad range of applications from nuclear reactor components to medical devices. FGMs composed of metal and ceramic are well-recognized for improving the properties of thermal-barrier systems, because cracking or delamination, usually detected in conventional multi-layer materials, are eliminated because of the smooth transition of composition and properties in the components used (Jha et al., 2013). In biomedical engineering, metal—ceramic porous FGM has been theoretically shown to have high potential in orthopedics such as the femoral component of knee prostheses, where the low modulus of elasticity (similar to bone modulus) is combined with outstanding anchorage to the bone on one side and high wear resistance on the other side (Bahraminasab et al., 2013).

One important issue in the successful design of hybrid materials is that it can usefully be biomimetic, such as considering human tissue structures or other biological assemblies like porous cancellous bone or graded cartilage structure. The other issue is the choice of constituents, volume fractions, configuration, and the way they can be connected to each other. This makes the design process a difficult task in which design and quality tools such as FEA, Design of Experiments, QFD, and MCDM analysis can be used in order to obtain the optimal material design. However, it should be noted that the development of hybrid materials needs innovative new processing routes that can directly control the desired variations of structure and properties.

1.11 Materials selection and sustainable products

The main aim of so-called "green" design or ecodesign, design for environment, and sustainable design is to manufacture products in a way that reduces the use of nonrenewable resources, uses less energy, does not directly or indirectly pollute the environment, and can be reused or recycled at the end of their useful life (Pfeifer,

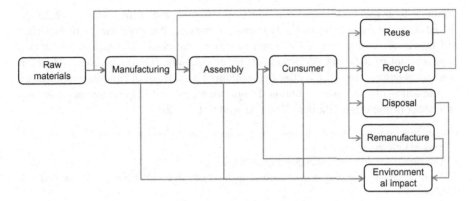

Figure 1.10 Raw materials, product life cycle, and recycling.

2009). The evaluation of a products' sustainability should consider the raw materials production, the fabrication process, energy consumption, transportation, and recycling. Therefore, one of the major opportunities for practicing sustainability is through material selection due to its potential impact on natural environmental systems. The options available at the end of a products' life are represented in Fig. 1.10 (Ishii et al., 1994).

The sustainability terms are defined as follows:

1. Recycling: The use of waste or a waste-derived material as a raw material for products, which may or may not be similar to the original.
2. Remanufacturing: Through certain refurbishing or restoration processes, some unserviceable products can regain the function and performances of the products which are similar to new ones (Yan and Gu, 1995).
3. Reuse: Further use of a waste product in its original form, such as the refilling of a previously discarded container (Yan and Gu, 1995).
4. Disposal: This refers to the elimination of the waste product without recovering any intrinsic value (Ishii et al., 1994).

In particular, a sustainable material selection means selecting materials that minimize environmental degradation over the whole life cycle of the material, from initial acquisition to eventual disposal or recycling (Djassemi, 2012).

Some examples of sustainable design of products influenced by materials occurring at the different stages of the product life cycle include the use of low toxic emission aqueous-based paint systems replacing solvent-based paint systems for motorcar bodies at the product manufacturing stage, the use of low energy consuming florescent light bulbs replacing incandescent light bulbs at the product use stage, and the use of recyclable or biodegradable beverage containers for the product disposal stage.

The processing of materials has significant impacts on the environment, including the use of water, land use patterns, undesirable emissions to air, water and land, and the consumption of other important environmental resources (Allwood et al., 2011). Therefore, materials selection in the product design process must be carried out carefully to achieve sustainable products. In this regard, Life Cycle Engineering (LCE) is one of the best methods. However, because of the large variety of materials and processes, considering LCE when selecting materials may require handling a large amount of data and performing many calculations (Djassemi, 2012). Hence it is reasonable to perform LCE after first obtaining a short list of candidate materials.

The following are some common design principles that help reduce environmental impact of materials (Pfeifer, 2009; Allwood et al., 2011):

- Use low-impact materials by choosing nontoxic materials or recycled materials that require little energy to process.
- Use manufacturing processes that require less energy.
- Make longer-lasting and better-functioning products that will have to be replaced less frequently.
- Design products for reuse and recycling.
- Design products with less material.
- Use materials that come from nearby (locally sourced).

There are more details about materials selection and the design and development of sustainable products in Ljungberg (2007).

1.12 Qualitative versus quantitative approaches to materials selection

A materials selection strategy has three main components (Ashby et al., 2004):

1. The formulation of constraints that must be satisfied if the material is to fill the desired function.
2. The formulation of a performance metric or value function to measure how well a material matches a set of requirements.
3. The use of a search procedure for exploring the solution-space to identify materials that meet the constraints, then ranking the materials by their ability to meet the requirements.

A structured set of guidelines to help select an optimal choice of materials when designing engineering products were developed by Edwards (2005). This is particularly beneficial for novice engineering designers but also serves as an important checklist for more experienced engineering designers.

For materials selection in which there are numerous different choices and many various criteria influencing the selection, it is difficult for engineering designers to evaluate qualitatively, current and new materials for a particular application, therefore an accurate approach is required (Dehghan-Manshadi et al., 2007). The conversion of qualitative information into quantitative information allows quantitative approaches to be used and provide the basis for direct comparisons to be made between each type of information. This approach is particularly useful at the conceptual design stage when information about materials is vague but becomes more problematic to apply further along the design process as the information about materials becomes more precise.

1.13   The role of computer-based materials selection and materials databases

The available set of materials are rapidly growing both in type and number (Roth et al., 1994). An engineering designer can only be expected to know a small part of the ever-growing body of knowledge on materials. Despite much of the materials information being available in standards, handbooks, and journal papers, actually finding this information is difficult and time-consuming and many potential materials might be overlooked. At one time, an engineering designer could rely on a handbook and personal experience to select the most appropriate materials for an application. Today, engineering designers are at risk of seeing their knowledge on materials easily go out of date, so are forced to look for systematic techniques for managing and analyzing engineering data on the growing array of materials (Roth et al., 1994; Ramalhete et al., 2010).

Fortunately, modern computer-based design tools, of which there are now many, support different aspects of the design decision-making process, but gaps still exist (Edwards and Deng, 2007). Digital libraries also are a learning tool to support the teaching of science and engineering fundamentals (Silva and Infante, 2015). A recent study has identified about 300 different software packages, databases, and website sources of information on materials (Ramalhete et al., 2010). This not only demonstrates the increasing number of materials' promoters and users but also the shortcomings of current systems for selecting materials. The current systems are either configured to provide decision support or are information based but their integration has been limited. Therefore, systems that mostly provide decision support leave the burden of considering the large number of alternatives on the decision maker (engineering designer in this case). Database systems (the majority of electronic sources of information on materials) contain a lot of data and information, but despite some using "search engines" and "user-friendly" interfaces, provide limited decision-making support when the number of criteria increases. The combination of decision theory and database technology has the potential to greatly improve the overall effectiveness of the materials selection process and therefore lead to better product design (Giachetti, 1998).

1.14   Summary and conclusions

The most common reasons for material substitution are: improving product service performance, meeting new legal requirements, reducing cost, and environmental issues. Therefore, materials selection/development is a permanent and ongoing obligation of engineers and designers involved in new product development. However, the selection of suitable materials for a specific purpose is an expensive time-consuming process because of the complex and large amount of information needed to be collected and analyzed. Therefore, designers are encouraged to use systematic techniques for managing and analyzing technical data on the growing array of materials. Product development involves a chain of decisions that include materials and design selection, process selection, and manufacturing equipment selection, that all benefit from MCDM, but the related decision support systems has received less attention. The chapters that follow describe the principal developments and use of MCDM techniques for selecting materials in the design of engineering products.

Review questions

1. Discuss the materials information needs at each stage of the design process (concept to detail) and its affect on decision-making.
2. Why is it necessary to be as thorough as possible and consider a large number of different materials in the early design stage?
3. Why are the consequences of making compromises in materials selection when designing a product?
4. What is the difference between material attribute identification and materials selection?
5. Why is it important to take into account manufacturing processes when selecting materials?
6. Discuss the issues involved when deciding to use a newly developed material in a product.
7. When selecting materials, why is it important to consider technical performance and user needs at the same time to achieve a successful design of product?
8. Discuss the differences between cost driven and performance driven product design with respect to materials utilization.
9. Explain the cost elements that should be considered by designers and discuss the uncertainty available on cost estimating.
10. Why is it essential to assess the consequences of any change in properties and the design of the whole product when substituting materials?
11. Describe the pros and cons of using multi-functional materials in the design of new products.
12. Discuss the influence of materials in the sustainable design of products occurring at the different stages of the product life cycle.
13. What are the advantages of using hybrid materials and why can sustainable design adversely affect their use?
14. Discuss the problems associated with incorporating qualitative information in a formalized materials selection process.
15. Why is it helpful to integrate design tools for facilitating improved decision-making when selecting materials?

References

Albiñana, J.C., Vila, C., 2012. A framework for concurrent material and process selection during conceptual product design stages. Mater. Des. 41, 433–446.

Allwood, J.M., Ashby, M.F., Gutowski, T.G., Worrell, E., 2011. Material efficiency: a white paper. Resour., Conserv. Recycl. 55, 362–381.

Ashby, M., 2013. Designing architectured materials. Scr. Mater. 68, 4–7.

Ashby, M., Shercliff, H., Cebon, D., 2013. Materials: Engineering, Science, Processing and Design. Butterworth-Heinemann, Oxford.

Ashby, M.F., 2005. Materials Selection in Mechanical Design. Butterworth-Heinemann, Oxford.

Ashby, M.F., Brechet, Y.J.M., Cebon, D., Salvo, L., 2004. Selection strategies for materials and processes. Mater. Des. 25, 51−67.

Asiedu, Y., Gu, P., 1998. Product life cycle cost analysis: state of the art review. Int. J. Prod. Res. 36, 883−908.

Bahraminasab, M., Sahari, B.B., Edwards, K.L., Farahmand, F., Hong, T.S., Naghibi, H., 2013. Material tailoring of the femoral component in a total knee replacement to reduce the problem of aseptic loosening. Mater. Des. 52, 441−451.

Blanch, R., Ferrer, I., Garcia-Romeu, M.L., 2011. A model to build manufacturing process chains during embodiment design phases. Int. J. Adv. Manuf. Technol. 59, 421−432.

Brechet, Y., Embury, J.D., 2013. Architectured materials: expanding materials space. Scr. Mater. 68, 1−3.

Brechet, Y., Bassetti, D., Landru, D., Salvo, L., 2001. Challenges in materials and process selection. Prog. Mater. Sci. 46, 407−428.

Callister, W.D., Rethwisch, D.G., 2007. Materials Science and Engineering: An Introduction. Wiley, New York, NY.

Chakladar, N.D., Das, R., Chakraborty, S., 2009. A digraph-based expert system for nontraditional machining processes selection. Int. J. Adv. Manuf. Technol. 43, 226−237.

Charles, J.A., Crane, F.A.A., Furness, J.A.G., 1997. Selection and Use of Engineering Materials. Butterworth-Heinemann, Oxford.

Chin, K.S., Chan, A., Yang, J.B., 2008. Development of a fuzzy FMEA based product design system. Int. J. Adv. Manuf. Technol. 36, 633−649.

Chiner, M., 1988. Planning of expert systems for materials selection. Mater. Des. 9, 195−203.

Dehghan-Manshadi, B., Mahmudi, H., Abedian, A., Mahmudi, R., 2007. A novel method for materials selection in mechanical design: combination of nonlinear normalization and a modified digital logic method. Mater. Des. 28, 8−15.

Dieter, G.E., 1983. Engineering Design: A Materials and Processing Approach. McGraw-Hill, Boston, MA.

Djassemi, M., 2012. A computer-aided approach to material selection and environmental auditing. J. Manuf. Technol. Manage. 23, 704−716.

Edwards, K.L., 2004. Strategic substitution of new materials for old: applications in automotive product development. Mater. Des. 25, 529−533.

Edwards, K.L., 2005. Selecting materials for optimum use in engineering components. Mater. Des. 26, 469−473.

Edwards, K.L., 2008. Towards an improved development process for new hip prostheses. Mater. Des. 29, 558−561.

Edwards, K.L., 2011. Materials influence on design: a decade of development. Mater. Des. 32, 1073−1080.

Edwards, K.L., Deng, Y.M., 2007. Supporting design decision-making when applying materials in combination. Mater. Des. 28, 1288−1297.

Farag, M.M., 1979. Materials and Process Selection in Engineering. Elsevier Science & Technology, London.

Farag, M.M., 2002. Quantitative methods of materials selection. In: Kutz, M. (Ed.), Handbook of Materials Selection. John Wiley & Sons, London.

Farag, M.M., 2008a. Materials and Process Selection for Engineering Design. CRC Press Taylor and Francis Group, London.

Farag, M.M., 2008b. Quantitative methods of materials substitution: application to automotive components. Mater. Des. 29, 374−380.

Giaccobi, S., Kromm, F.X., Wargnier, H., Danis, M., 2009. Filtration in materials selection and multi-materials design. Mater. Des. 31, 1842−1847.

Giachetti, R.E., 1998. A decision support system for material and manufacturing process selection. J. Intell. Manuf. 9, 265−276.

Henriques, E., Pecas, P., Silva, A., Leite, M., 2014. On the influence of material selection decisions on second order cost factors. Technology and Manufacturing Process Selection. Springer, London.

Ishii, K., Eubanks, C.F., Di Marco, P., 1994. Design for product retirement and material life-cycle. Mater. Des. 15, 225−233.

Jahan, A., 2012. Material selection in biomedical applications: comparing the comprehensive VIKOR and goal programming models. Int. J. Mater. Struct. Integrity. 6, 230−240.

Jahan, A., Edwards, K.L., 2013. Weighting of dependent and target-based criteria for optimal decision-making in materials selection process: biomedical applications. Mater. Des. 49, 1000−1008.

Jahan, A., Ismail, M.Y., Sapuan, S.M., Mustapha, F., 2010. Material screening and choosing methods—a review. Mater. Des. 31, 696−705.

Jalham, I.S., 2006. Decision-making integrated information technology (IIT) approach for material selection. Int. J. Comput. Appl. Technol. 25, 65−71.

Jha, D., Kant, T., Singh, R., 2013. A critical review of recent research on functionally graded plates. Composite Struct. 96, 833−849.

Karana, E., 2006. Intangible characteristics of materials in industrial design. The International Conference on Design and Emotion.

Karana, E., Hekkert, P., Kandachar, P., 2008. Material considerations in product design: a survey on crucial material aspects used by product designers. Mater. Des. 29, 1081−1089.

Kazemzadeh, R.B., Behzadian, M., Aghdasi, M., Albadvi, A., 2009. Integration of marketing research techniques into house of quality and product family design. Int. J. Adv. Manuf. Technol. 41, 1019−1033.

Khan, A.S., Azam, M.T., Khan, M., Mian, S.A., Rehman, I.U., 2015. An update on glass fiber dental restorative composites: a systematic review. Mater. Sci. Eng. C. 47, 26−39.

Kusiak, A., 1993. Concurrent Engineering: Automation, Tools, and Techniques. John Wiley & Sons.

Layer, A., Brinke, E.T., Houten, F.V., Kals, H., Haasis, S., 2002. Recent and future trends in cost estimation. Int. J. Comput. Integr. Manuf. 15, 499−510.

Ljungberg, L.Y., 2007. Materials selection and design for development of sustainable products. Mater. Des. 28, 466−479.

Ljungberg, L.Y., Edwards, K.L., 2003. Design, materials selection and marketing of successful products. Mater. Des. 24, 519−529.

Lovatt, A.M., Shercliff, H.R., 1998. Manufacturing process selection in engineering design. Part 1: the role of process selection. Mater. Des. 19, 205−215.

Mayyas, A., Shen, Q., 2011. Using quality function deployment and analytical hierarchy process for material selection of body-in-white. Mater. Des. 32, 2771−2782.

Michael, P., 2009. Selecting materials. Materials Enabled Designs. Butterworth-Heinemann, Boston, MA (Chapter 3).

Noble, J.S., Tanchoco, J., 1990. Concurrent design and economic justification in developing a product. Int. J. Prod. Res. 28, 1225–1238.

Normung, D.I.F., 1989. DIN 32992 Teil 1: Kosteninformationen, Berechnungsgrundlagen, Kalkulationsarten Undverfahren. Deutsches Institut für Normung, Berlin.

Pfeifer, M., 2009. Design requirements. Materials Enabled Designs. Butterworth-Heinemann, Boston, MA (Chapter 2).

Ramalhete, P.S., Senos, A.M.R., Aguiar, C., 2010. Digital tools for material selection in product design. Mater. Des. 31, 2275–2287.

Roth, R., Field, F., Clark, J., 1994. Materials selection and multi-attribute utility analysis. J. Comput. Aided Mater. Des. 1, 325–342.

Schifferstein, H.N.J., Hekkert, P. (Eds.), 2007. Product Experience. Elsevier.

Silva, A., Infante, V., 2015. The role of digital libraries in teaching materials science and engineering. Handbook of Research on Recent Developments in Materials Science and Corrosion Engineering Education, Hershey, PA, USA, IGI Global.

Thurston, D.L., Locascio, A., 1994. Decision theory for design economics. Eng. Econ. 40, 41–71.

Ulrich, K.T., Eppinger, S.D., 2004. Product Design and Development. McGraw-Hill/Irwin, Boston, MA.

van Kesteren, I., 2008. Selecting Materials in Product Design. Faculty of Industrial Design Engineering, Delft, Delft University of Technology.

van Kesteren, I.E.H., Kandachar, P.V., Stappers, P.J., 2006. Activities in selecting materials by product designers. In: Proceedings of the International Conference on Advanced Design and Manufacture, Harbin, China.

Yan, X., Gu, P., 1995. Assembly/disassembly sequence planning for life-cycle cost estimation. Manuf. Sci. Eng., ASME. MED-2 (2)/Mh-3 (2), 935–956.

Zarandi, M.H.F., Mansour, S., Hosseinijou, S.A., Avazbeigi, M., 2011. A material selection methodology and expert system for sustainable product design. Int. J. Adv. Manuf. Technol. 57, 885–903.

Materials selection in the context of design problem-solving

2

Learning Aims

The overall aim of this chapter is to gain an overview of systematic and creative design tools and principles in materials design/selection. After carefully studying this chapter you should be able to understand:

- the basic concept of inventive problem-solving, and its role in design
- the relationship between decision support system in materials selection, and trends in lean design principles
- the conceptual quality function deployment (QFD) models for materials selection
- the relationship between QFD and multi-criteria decision making (MCDM) for material design/selection.

2.1 Creative thinking and inventive problem-solving (TRIZ)

Projects of all kinds often reach a point where the next stage is unclear. To understand what to do, the project team needs to be creative. The creativity tools commonly used are brainstorming (the most popular), to synectics, removing mental blocks, and morphological charts, which depend on the knowledge and perception of the team members. TRIZ is a problem-solving method based on logic and data, not intuition, and speeds up the project team's ability to solve problems creatively. TRIZ is the (Russian) acronym for the "Theory of Inventive Problem Solving." TRIZ is an international science of creativity that relies on the study of patterns of problems and solutions. To discover the patterns that predict breakthrough solutions to problems, more than three million patents have been examined. The development of TRIZ started with the hypothesis that there are general principles of creativity that are the basis for creative innovations that advance technology. If the related principles could be codified, then these can be taught to people to make the process of creativity more predictable. In other words, somebody somewhere has already solved a very similar problem. Creativity now becomes finding that solution and adjusting it to the particular problem being solved (Fig. 2.1). Since TRIZ is built on a large database of thousands of different patents, principles, and contradictions, the problem-solving method has been implemented as a computer software application to help engineers and designers, with minimal training, rapidly achieve meaningful results.

Multi-criteria Decision Analysis. DOI: http://dx.doi.org/10.1016/B978-0-08-100536-1.00002-3

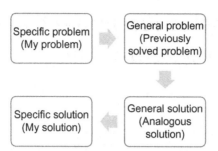

Figure 2.1 The TRIZ problem-solving method.

Design and development approaches, such as quality function deployment (QFD), show us "what" to solve but not always "how" to solve the technology restrictions that can occur. For instance, when designing a mechanical dampener, the search for a design solution might be limited to the use of rubber-based materials. However, creating a magnetic field could be a more efficient design solution. Given that this might be outside the knowledge of a particular designer, such an alternative solution would not necessarily be realized. By means of TRIZ, it is possible for designers to readily explore different design solutions in other fields, or confirm the initial solution was the best.

2.2   Comparison between creative and systematic design methods

Creativity is an integral and essential part of the engineering design process (Howard et al., 2008). In industry, however, creativity does not necessarily lead to success but without it long-term failure is a near certainty. Understanding how the engineering design process is related to the creative problem-solving process will give insight into where and when important resources should be focused in order to enhance creative performance and also the resulting "quality" of the product designed (Howard et al., 2008). Fig. 2.2 summarizes the main phases of the creative and systematic design processes. A systematic design methodology ensures that no important possibilities are missed, and reduces the time-to-market by completing the new product development (NPD) process as soon as possible. It reforms the NPD process by encouraging concurrent NPD. A systematic design methodology can also help manage the required levels of creativity. On the negative side, however, by prescribing behavior and decision rules, it reduces the possibility of generating radical or unconventional ideas, and does not make NPD teams strive for maximum creativity. It can also reduce the team's creative performance when managed by a person unaware of the need to balance and apply the principles properly (Leenders et al., 2007). It is recognized that the engineering design process models are generally considered to be poor with regards to representing creative processes

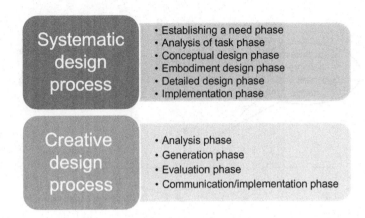

Figure 2.2 The main phases of the systematic and creative design processes.

(Howard et al., 2008). In business organizations today, creativity and innovation are less often the product of individual genius and more often the outcome of processes in teams (Leenders et al., 2007).

2.3 Identifying the needs and specifying the property requirements for materials

Customers do know what they desire but might not be expert at describing their needs properly. Their needs come from a variety of sources with different motivations, and in simple terms are a mixture of quantitative, qualitative, and intangible factors (Jack, 2013). Examples include: Inventors (a perceived need), Entrepreneurs (a project essential to establishing a new business), Sponsors/customers (a group that comes with a previously established need; they may also provide specifications), Yourself (a self-identified project that has some value to solve your own needs), Social (a humanitarian project motivated by helping people in need), and Competition (a design objective constrained by contest rules).

Design specifications are the minimum acceptance criteria and are developed from customer needs. Without clear customer needs, every design solution could be either accepted or rejected on a whim (Jack, 2013). The Kano model concept is valuable for setting an expectation for design. Mixing this concept with QFD matrices provides important insights into the dynamics of the customer, and leads to distortion in the customer weighting of product characteristics. QFD is a formal method for incorporating consumer needs into the design process and is briefly explained in the following section. Kano's model focuses on differentiating product features, as opposed to focusing initially on customer needs. The model involves the two

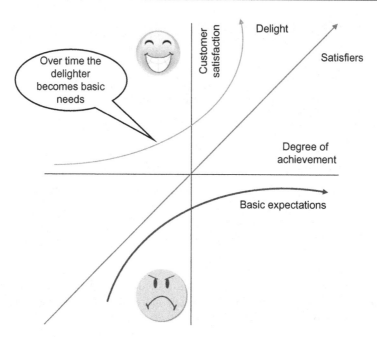

Figure 2.3 The Kano model.

dimensions of achievement and satisfaction (Fig. 2.3). The three levels of customer expectations include "Expected Needs," "Normal Needs," and "Exciting Needs":

- Expected Needs (Basic expectations): These expectations are also known as the "dissatis-fiers" because by themselves they cannot fully satisfy a customer (eg, attributes relative to safety). However, failure to provide these basic expectations will cause dissatisfaction.
- Normal Needs (Satisfiers): These are the qualities, attributes, and characteristics that keep a supplier in the market. These next higher level expectations are known as the "satis-fiers" (eg, lower price or guaranteed services).
- Exciting Needs (Delighters): These are features and properties that make a supplier a leader in the market. These expectations are also known as the "delighters" or "exciters" because they go well beyond anything the customer might imagine and ask for. Their absence does nothing to adversely affect a possible sale, but their presence improves the likelihood of a purchase. Over time, "exciting needs" become "expected needs" (the mobile telephone market is a good example).

The main steps of the Kano model analysis are:

1. Brainstorm all of the possible features of your product and everything you can do to please your customers.
2. Classify these as "Threshold" or "Expected Needs," "Performance" or "Normal Needs," "Excitement" and "Not Relevant." Where possible, get your customers to do the classifi-cation for you.
3. Make sure your product has all the appropriate Threshold Attributes. The product will not be successful if these attributes are not present.
4. Where possible, remove the attributes that are "Not Relevant."

5. Select appropriate Performance Attributes so that you can deliver a product at an appropriate price the customer is prepared to pay.
6. Look at the Excitement Attributes and think how you can build some of these into your product. The market punishes too little or too much change, but rewards incremental improvements.

The successful design of an engineering component is integral to satisfying the functional and customer specified requirements for the overall product it forms a part, utilizing material properties and capabilities of suitable manufacturing processes (Zarandi et al., 2011). Fig. 2.4 shows a flowchart of materials selection and continuous improvement in product development. There is the option to invent a new material, but this involves cost, time, and risk. QFD helps marketing and design teams incorporate the "voice of the customer" in product designs, increasing the likelihood that the final product will successfully satisfy the customer's needs. Fundamentally, the strategy for selecting materials is determined by the need to satisfy design requirements.

2.4   Determining the characteristics of materials using QFD

The success of a company depends largely on meeting the real and hidden needs of its customers (Prasad and Chakraborty, 2013). The relative preferences of these requirements have a direct influence on the product design and this is normally imposed through relating design parameters to material properties. These parameters could be either a function of a single material property or a combination of two or more properties (material indices). The extraction of all the relevant material indices needs a broad engineering knowledge and this sometimes makes it difficult for practitioners, particularly novices, to employ the selection methods appropriately. In fact, even mistakenly missing a material index may adversely affect the results (Kasaei et al., 2014). Fig. 2.5 shows the relationship between some functional requirements and material and design characteristics. Although a large amount of multi-criteria decision making (MCDM)-based research has already been carried out on materials selection, it has been observed that the mathematical approaches do not fully take into account the "voice of the customers" and hence their requirements in the final design of products (Prasad and Chakraborty, 2013).

Fig. 2.6 shows an example of mapping performance requirements of a component relating the main characteristics of materials and processing requirements for the design of a bicycle.

2.4.1   A brief review QFD

QFD is a systematic method to help identify customers' needs for designing a product (or service) in a way that it considers the customers' needs first time (Prasad and Chakraborty, 2013). The QFD approach uses a matrix format that looks like a house and, hence, it is also known as the "House of Quality" (HoQ). The method

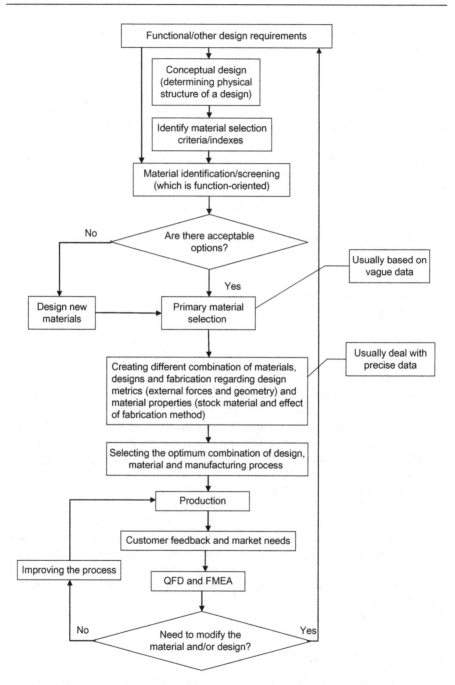

Figure 2.4 Specifying the property requirements for materials in context of continuous improvement in product development (Jahan and Edwards, 2013).

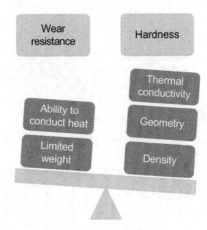

Figure 2.5 Example of mapping between material/design selection parameters and functional requirements.

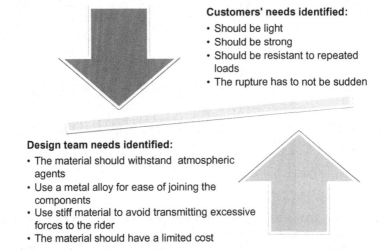

Customers' needs identified:
- Should be light
- Should be strong
- Should be resistant to repeated loads
- The rupture has to not be sudden

Design team needs identified:
- The material should withstand atmospheric agents
- Use a metal alloy for ease of joining the components
- Use stiff material to avoid transmitting excessive forces to the rider
- The material should have a limited cost

Figure 2.6 Customers' needs in the design of a bicycle and complementary requirements detected by the design team related to candidate materials (Cavallini et al., 2013a).

aids in making cognizant decisions about the voice of customers. The aim of QFD is to translate objective and even subjective quality criteria into objective quality criteria that can be quantified and measured. It is a complementary approach for indicating how and where priorities are to be assigned in product development. For implementing QFD, three main steps are taken:

1. Prioritize spoken and unspoken customer wants or needs.
2. Translate the identified needs into technical terms and specifications.
3. Build and deliver a quality product or service by concentrating everybody towards the customer satisfaction.

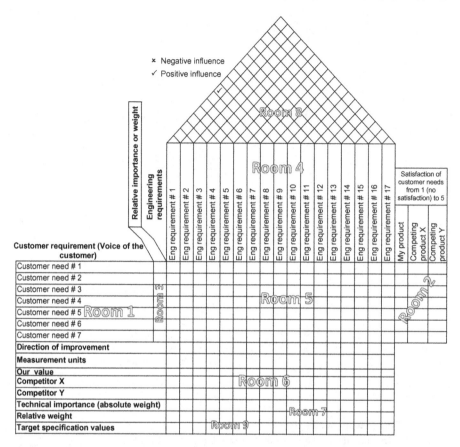

Figure 2.7 House of Quality.

The HoQ (also known as the product planning matrix) is a primary tool used in QFD and usually comprises of nine different sections, or so-called Rooms. The Rooms in the HoQ must be completed or filled step-by-step as shown in Fig. 2.7.

Room 1 contains the customer requirements (CRs) (the "Whats") and *Room 2* is used to carry out competitive analysis of the company's and competitors' products with regard to the CRs. Furthermore, marketing knowledge is added to *Room 2*, in order to distinguish between some properties of the CRs. Both *Rooms 1* and 2 need knowledge of the market place, hence are best filled by the Marketing Department. *Room 3* encompasses the relative importance of the CRs and can be judged according to a priority scale developed as 1—not important, 2—important, 3—much more important, 4—very important, and 5—most important. In this room, the CRs are quantified and ranked in order of their importance. Following this procedure, the requirements demanding more improvement are considered as prime selling points by Marketing, thus obtain higher weighting. These customers' requirements might be beneficial (higher values are desired) or nonbeneficial (lower values are preferred). The beneficial or nonbeneficial nature of the CRs is dictated by the value of the corresponding improvement driver ($+1$ for beneficial attribute and -1 for nonbeneficial attribute). Engineers record their "voice" in *Room 4* in the form of engineering requirements (the "Hows"). The technical terms may also be beneficial or nonbeneficial, or target-based (Cavallini et al., 2013b).

Room 5 is an important room in the house, which records in each cell the influence that an engineering requirement has on gaining a corresponding CR. To fill up the HoQ matrix and to develop the interrelationship matrix indicating the relations between the customers' needs and technical requirements (the "Whats" and "Hows"), an appropriate scale is set for assigning the relative importance as 1—very weak, 3—weak, 5—moderate, 7—strong, and 9—very strong. The interrelationship matrix is marked by use of symbols or numbers according to the contribution made by each customer's requirement and technical requirement.

A technical comparison between the company's and competitors' products conducted by the designers is recorded in *Room 6*. In *Room 7*, the technical requirements are quantified and ranked in order of their importance. When the HoQ matrix is completely filled by the necessary data, the weighting of each technical requirement is calculated based on Eq. (2.1):

$$W_j = \sum_{i=1}^{n} R_{ij} \times C_i \quad j = 1, 2, \ldots, m \tag{2.1}$$

where W_j is the absolute weighting for the jth technical requirement, n is the number of customers' requirements, C_i is the priority assigned to the ith CR, m is the number of engineering characteristics, and R_{ij} is the weighting assigned to the relationship between the jth technical requirement and the ith CR. The relationship between technical requirements is shown in *Room 8*, which supports the product design. It is also known as a roof matrix. The analysis related to the roof of quality is improved when technical/engineering characteristics influence each other in asymmetric ways and their mutual influence varies in relation to different CRs (Reich and Levy, 2004). The end result of the QFD is *Room 9*, which records the target set manually by the development team after taking into account the weighting, cost, and technical difficulty as well as the decision trade-offs from *Room 8*. It is worth mentioning that QFD and TRIZ can complement each other. TRIZ can help to eliminate contradictions discovered by the roof of the HoQ and on determining target values as well as developing new concepts for materials and design. Fig. 2.8 shows schematically a breakdown of the general four-phase QFD process model.

2.4.2 Conceptual QFD model for materials selection

Several mathematical approaches have already been proposed to aid designers in selecting the most appropriate materials for diverse engineering applications. These methods usually employ criteria weighting values in their computations, which are usually based on the subjective judgments of the designers (Prasad and

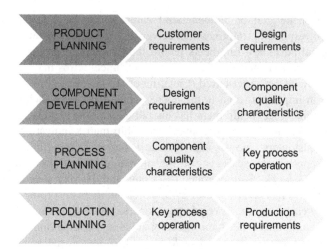

Figure 2.8 The QFD process.

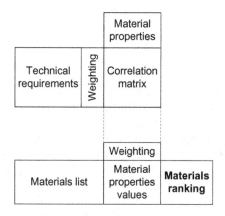

Figure 2.9 Conceptual model of two-phase QFD for materials selection.

Chakraborty, 2013). The approach for materials selection using QFD has two main steps (Scalice et al., 2012; Prasad and Chakraborty, 2013):

1. development of the HoQ matrix
2. construction of the material score matrix for the alternatives considered.

Fig. 2.9 shows the two-phase QFD model for materials selection, and Fig. 2.10 demonstrates an example of the correlation between technical requirements and material properties. The conceptual QFD-based materials selection model can be used for each component of a product, or for some systems or subsystems, when it is feasible. It relates the design or technical requirements for each product with the material properties and then selects the most suitable material based on these material properties.

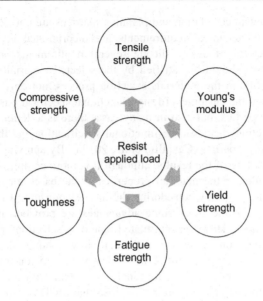

Figure 2.10 Example of relationship between technical requirement and material properties.

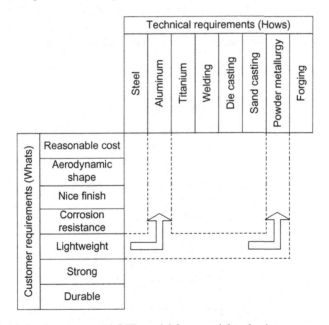

Figure 2.11 Alternative conceptual QFD model for materials selection.

The HoQ matrix may take various forms depending on the type of problem to be solved. Fig. 2.11 demonstrates an alternative conceptual QFD model for materials selection that might be useful in some situations. However, the aforementioned method is more acceptable in the literature.

Where the technical correlation and planning matrices are not taken into account, only the prioritized technical requirements are incorporated at the base of the matrix. The objectives of design dictate several requirements on every product, which some or most should be satisfied by the selected material(s) for the design (Kasaei et al., 2014). In the materials selection process, not only it is not easy to identify the voice of the customers (design functional specifications) that are related to the material properties, but also in many cases there is a deficiency in the mapping of the connections between the functional specification and the physical characteristics/material properties (Cavallini et al., 2013a). By applying material indices it assures the generality of the results, although it is more complicated compared to the methods considering the material properties as the bases for selection (Kasaei et al., 2014). Furthermore, by considering the interrelationship of the material indices through the use of the HoQ, some advantages are provided in comparison to other existing methods. However, correlated materials selection criteria sometimes produce an incomplete understanding of the optimal weighting.

The HoQ can be used as an initial aid to deal with decision trade-offs, and can be coupled with multi-criteria decision analysis to deal with the design selection process strategically. It can therefore be assumed that MCDM will improve the second phase of the four-phase QFD model (see Fig. 2.8), and increase the precision of process planning. Fig. 2.12 shows the process of translating customer needs to determine weightings for materials and design selection criteria as well as design target values. These outputs are vital to the success of any new product.

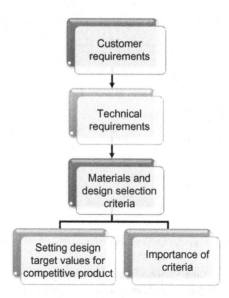

Figure 2.12 Process of translating customer needs into set target values and weighting of criteria.

2.5 Utilization of a "Lean" design principles approach to materials selection process

The term "Lean" (Haque and James-Moore, 2004) refers to the concepts of "Lean thinking," or "Lean principles," broadly made popular within manufacturing operations. Successes in lean manufacture have guided researchers and practitioners to study extending "Lean" principles in various parts of the engineering enterprise, including Product and Process Development (PPD) (Khan et al., 2013). The outstanding commercial performance of Toyota Motor Corporation is good evidence of the benefits brought by adopting "Lean" practices. Interestingly, most western manufacturers are focusing their "Lean" initiatives at operations with few attempts to adopt "Lean" in design-related activities. However, achieving such improvements is often difficult and it is clear that the benefits of "Lean" cannot be fully realized simply by adopting a few miscellaneous tools and techniques.

It appears to be a noteworthy opportunity to take advantage from the adoption of "Lean" in engineering design (Baines et al., 2006). Generating the right information in the right place and at the right time creates value for "Lean" in product design, hence the standardization of knowledge/information management processes help the adoption of PPD that it is yet to be defined (Baines et al., 2006). The definition of "Lean" has altered during the years, from a philosophy for waste reduction to the emerging view as value creation (Fig. 2.13). "Lean" is about creating more value for customers by removing activities which are considered wasteful, but value in the product design process requires precise definition as it is not necessarily the same as value in manufacturing operations.

The five principles of "Lean" thinking are shown in Fig. 2.14. These principles have to be tailored for the new product introduction (NPI)/engineering processes, with the following statements created (Haque, 2003):

1. Identify the value accurately from the view point of the final customer, as well as the internal and external stakeholders, in terms of particular products, information, and services, with particular capabilities accessible at a specific price and time.

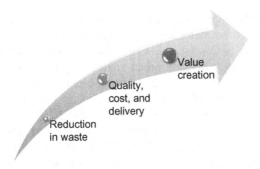

Figure 2.13 The changing definition of "Lean" (emphasis from manufacturing to design).

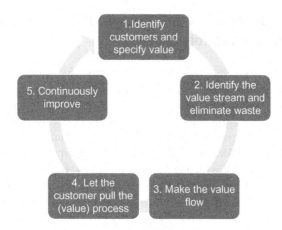

Figure 2.14 The five "Lean" principles for maximizing value.

2. Make a hierarchical model of the NPI value stream which delivers the specified value and discards nonvalue adding processes and activities by evaluating that value stream.
3. Design and accomplish the desired NPI value stream that makes the remaining value added steps flow.
4. Plan a program in which activities, their workload, and objectives are considered depending on the needs of the customer (downstream activities).
5. Continually determine ways to have increased value provision, reduce the costs of nonvalue adding but required activities, and remove successive layers of waste, since they are uncovered in current activities.

The "Lean" principles suggested should not be limited to NPD only, but should also concern all aspects of a company's innovation efforts related to providing a satisfying customer experience (Gudem et al., 2014).

2.6 Summary and conclusions

The strategy for selecting materials is determined by the need to satisfy design requirements that are normally imposed through relating design parameters to material properties or performance indices. However, the extraction of all the relevant material parameters needs a broad engineering knowledge and even omitting a selection criterion may adversely affect the outcomes. QFD can be used for classifying and translating the design requirements into materials and design selection criteria, and can be coupled with MCDM strategically. Although "Lean" principles have been applied efficiently to manufacturing and operations processes in many companies, their use in product design, specifically the materials selection/design process, has been inadequate and is an emerging topic that can also benefit from MCDM.

Review questions

1. Describe the relationships between creative problem-solving and the design process.
2. Explain the role of the Kano model in the analysis of customer needs.
3. How do patents help in the generation of new design ideas?
4. List four CRs each for a: mountain bicycle, racing bicycle, and children's bicycle, and discuss the differences.
5. Clarify the place of materials selection in the four-phase QFD model.
6. Suppose there is a blank column (or row) in the relationship matrix (the "Whats" and "Hows," respectively). What conclusions can you draw?
7. Discuss the possible advantages of "Lean" thinking in NPD.

References

Baines, T., Lightfoot, H., Williams, G.M., Greenough, R., 2006. State-of-the-art in lean design engineering: a literature review on white collar lean. Proc. Inst. Mech. Eng. B: J. Eng. Manuf. 220, 1539−1547.

Cavallini, C., Citti, P., Costanzo, L., Giorgetti, A., 2013a. An axiomatic approach to managing the information content in QFD: applications in material selection. In: The Seventh International Conference on Axiomatic Design, Worcester.

Cavallini, C., Giorgetti, A., Citti, P., Nicolaie, F., 2013b. Integral aided method for material selection based on quality function deployment and comprehensive VIKOR algorithm. Mater. Des. 47, 27−34.

Gudem, M., Steinert, M., Welo, T., 2014. From lean product development to lean innovation: searching for a more valid approach for promoting utilitarian and emotional value. Int. J. Innov. Technol. Manage. 11, 1450008.

Haque, B., 2003. Lean engineering in the aerospace industry. Proc. Inst. Mech. Eng. B: J. Eng. Manuf. 217, 1409−1420.

Haque, B., James-Moore, M., 2004. Applying lean thinking to new product introduction. J. Eng. Des. 15, 1−31.

Howard, T.J., Culley, S.J., Dekoninck, E., 2008. Describing the creative design process by the integration of engineering design and cognitive psychology literature. Des. Stud. 29, 160−180.

Jack, H., 2013. Needs identification and specifications. Engineering Design, Planning, and Management. Academic Press, Boston, MA (Chapter 2).

Jahan, A., Edwards, K.L., 2013. Weighting of dependent and target-based criteria for optimal decision-making in materials selection process: biomedical applications. Mater. Des. 49, 1000−1008.

Kasaei, A., Abedian, A., Milani, A., 2014. An application of quality function deployment method in engineering materials selection. Mater. Des. 55, 912−920.

Khan, M.S., Al-Ashaab, A., Shehab, E., Haque, B., Ewers, P., Sorli, M., et al., 2013. Towards lean product and process development. Int. J. Comput. Integr. Manuf. 26, 1105−1116.

Leenders, R.T.A., Van Engelen, J.M., Kratzer, J., 2007. Systematic design methods and the creative performance of new product teams: do they contradict or complement each other? J. Prod. Innov. Manage. 24, 166−179.

Prasad, K., Chakraborty, S., 2013. A quality function deployment-based model for materials selection. Mater. Des. 49, 525—535.

Reich, Y., Levy, E., 2004. Managing product design quality under resource constraints. Int. J. Prod. Res. 42, 2555—2572.

Scalice, R.G.K., Brascher, G.C., Becker, D., 2012. A knowledge-based material selector using quality function deployment principles. Prod. Manage. Dev. 10, 23—32.

Zarandi, M.H.F., Mansour, S., Hosseinijou, S.A., Avazbeigi, M., 2011. A material selection methodology and expert system for sustainable product design. Int. J. Adv. Manuf. Technol. 57, 885—903.

Screening of materials

3

Learning Aims

The overall aim of this chapter is to gain an overview of screening methods used for materials selection. After carefully studying this chapter you should be able to understand:

- the principles of current formalized screening methods
- the significance of the questionnaire/guideline method for material screening
- the importance of updating material databases
- the basics of the chart-based method (the "Ashby" approach) for material screening
- the limitations of screening methods used for the final selection of material.

3.1 Justification for an initial screening process

It is normal in the early stages of the design process for the number of materials under consideration to be very large. To limit the total effort required, it is important to find out the types of materials that are possible for use in the design of a specific component. During the screening stage (preselection), the types of materials that are possible to utilize are identified. At the same time design configurations are determined and manufacturing methods are considered. As the design progresses the number of materials reduces and the data becomes more refined until eventually a candidate material is selected and used (Edwards and Deng, 2007). In order to screen materials effectively it is necessary to apply materials selection methods.

The output from the screening process is a set of candidate materials that can be taken to the next stage of the design process and more detailed evaluation. The materials selection process is further complicated by the need to also consider manufacturing processes when designing components. The number of different materials available is considerable, with new materials also being developed all the time. As a consequence it is all too easy for product designers not to be aware of all material possibilities and their associated processing limitations. Also, databases and materials selection tools, the means of accessing materials information, are constantly evolving. Product designers therefore need to keep themselves up to date with developments in materials selection methods, materials databases, and computer-based design tools to help avoid missing suitable materials, especially new materials. The use of new materials has risks but may provide important performance or economic advantages. This justifies the need to thoroughly screen materials, existing and new materials where possible, when designing components for new or improved products.

Multi-criteria Decision Analysis. DOI: http://dx.doi.org/10.1016/B978-0-08-100536-1.00003-5

3.2   Introduction to the use of material attributes

Only occasionally will the exact grade of a material be specified by customers. Even then the designer must understand the material used to be able to design the product, because a range of properties is possible for a particular material. This is created by variability in the material itself and by the processing, for which several options may be available. Product analysis enables an understanding of the significance of materials, processing, economics, and esthetics before any product can be manufactured. A consideration of the following questions can help to build a suitable design specification (Lovatt et al., 2000):

- What is the function of each component in the product, and how do they work?
- What requirements should each component possess to perform well (eg, electrical, mechanical, aesthetic, ergonomic, etc.)?
- What manufacturing processes are employed to manufacture each component and why?
- What type of material is used to make each component and why?
- Are there any other materials or design options available and is it possible to propose improvements?

In the case of a bicycle for instance, its function depends partly on its type (Racing, Touring, Mountain, Commuter, or Children's bicycle). As well as considering the necessary mechanical/structural requirements, it is also important to consider the ergonomic aspects, how the bicycle design has been made user-friendly, and any marketing issues because these all influence the later design decision-making. The following questions can help determine the design requirements for each type of bicycle:

- How does the bicycle operate?
- How is the function related to the type of bicycle?
- How will the bicycle be manufactured to be minimally maintained?
- What should be the appearance of the bicycle (eg, color)?
- What should the bicycle cost?
- How has the bicycle been made comfortable to ride?
- How do the different components of the bicycle operate and interact with each other?

Given the specification of requirements for each component, the material properties that are important can be highlighted. As an example, for a requirement of "must support loads without breaking," the significant material property is "strength." Quality function deployment can also be used as a tool for translating customer needs into design requirements.

Designers and engineers have to take into account a large number of different factors when selecting materials. Karana et al. (2008) showed that when designers select a material they must consider fulfilling the following quantitative requirements:

1. mechanical properties (eg, Young's modulus, ultimate strength, yield strength, elasticity, fatigue, creep resistance, ductility, hardness, toughness, etc.)
2. fabrication requirements (eg, machining ability, formability, welding ability, casting ability, heat treatability, etc.)

3. economic requirements

4. maintenance.

5. thermal and radiation properties (eg, specific heat, conductivity, expansivity, diffusivity, transmissivity, reflectivity, emissivity, etc.)

6. corrosion/oxidation

7. wear

8. physical properties (eg, crystal structure, density, melting point, vapor pressure, viscosity, porosity, permeability, reflectivity, transparency, optical properties, dimensional stability, etc.)

9. chemical properties.

10. electrical properties (eg, resistivity, permittivity, dielectric strength, etc.)

11. acoustical properties

12. optical properties

13. dimensional properties

14. business issues

15. life of component factors

16. availability

17. esthetic attributes

18. ecoproperties (eg, recyclability, sustainability, embodied energy, CO_2 "footprint", etc.).

Besides these requirements, recently subjective and qualitative user-interaction aspects, such as appearance, perceptions, and emotions, are being considered in materials selection (Ashby and Johnson, 2013; Crilly et al., 2004; van Kesteren et al., 2007a; Ashby et al., 2013). The user-interaction aspects are those that have an effect on the usability and personality of a product and are formulated as sensorial material properties. As a consequence, materials acquire different meanings in different products. Recent research by Karana et al. (2009) showed the association of materials with certain subjective meanings such as aggressive, nostalgic, professional, sexy, and toy-like. In competing products where technical performance is roughly equal, subjective meaning is important and this could make the difference between the success and failure of a product in the market place. However, taking into account both quantitative and qualitative properties simultaneously are difficult when selecting materials and at present rely heavily on the experience of the product designer for the latter.

3.3 Material performance indices and critical material aspects used by product designers

Materials play a significant role in design, that is, material attributes (properties) define (or limit) performance. Most products need to satisfy certain performance targets, which impacts on the design of the components making up the product, and are determined by considering the design specification. For example, the product (and its components) might need to be cheap, or stiff, or strong, or light, or perhaps some or all of these requirements. Each of these performance requirements will influence materials selection. If the product needs to be light, Lead would not be

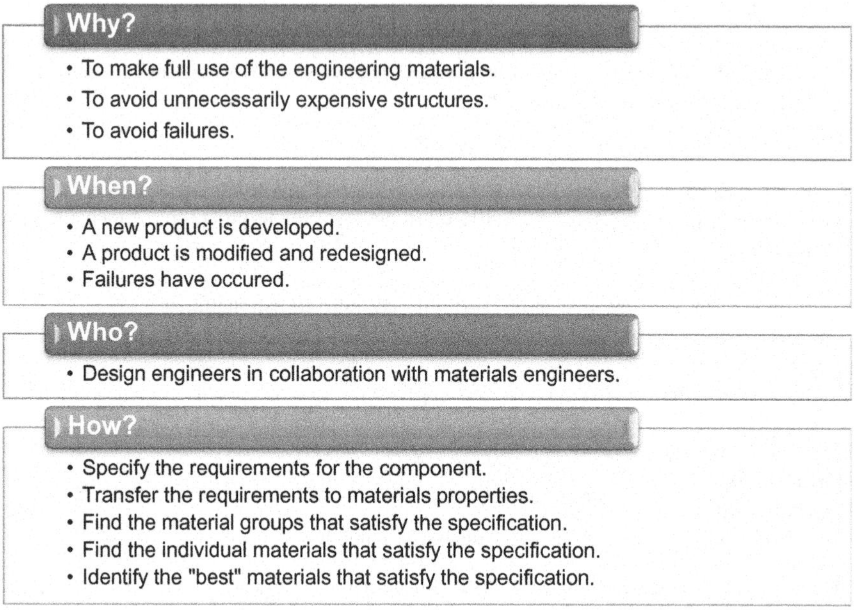

Why?
- To make full use of the engineering materials.
- To avoid unnecessarily expensive structures.
- To avoid failures.

When?
- A new product is developed.
- A product is modified and redesigned.
- Failures have occured.

Who?
- Design engineers in collaboration with materials engineers.

How?
- Specify the requirements for the component.
- Transfer the requirements to materials properties.
- Find the material groups that satisfy the specification.
- Find the individual materials that satisfy the specification.
- Identify the "best" materials that satisfy the specification.

Figure 3.1 Materials selection basics (Sandstrom, 1994).

chosen and if it needs to be stiff, rubber would not be chosen, unless there are compelling reasons. There are circumstances when a material might be used that is not the best available but overall satisfies all the design requirements, and this is usually down to cost. Also, materials adopted in the past may be deemed unsuitable today because of better alternative materials now available, eg, polymers replacing wood for use in window frames. The basics of materials selection are shown in Fig. 3.1.

Materials selection problems can be summarized into two distinct categories (Deng and Edwards, 2007):

1. Materials selection based on the material properties (mechanical, physical, electrical, thermal, chemical, etc.).
2. Materials selection based on the design requirements, where the material properties are coupled with those of the physical structure and the relevant structural properties of the component.

In contemporary applied scientific research, such as engineering, the performance of materials is normally evaluated against multiple criteria, rather than considering a single criterion, although the latter is still important. For example, in aircraft structures or airframes (eg, fuselage, wings, tail, control surfaces, etc.), the materials used must be light, stiff, and strong, hence the prevalent use of high-strength aluminum alloys but also increasing use of fiber-reinforced polymer matrix composites (eg, carbon fiber/epoxy laminates), where the cost is justified. The use of carbon fiber composite materials also facilitates greater design freedom, eg, material property tailoring, complex geometry, parts consolidation, etc., which is not possible with metals. It is now common to see both metallic and polymeric materials being combined in the same structure, thereby simultaneously utilizing the advantages of both materials, to achieve the desired performance (Jahan and Edwards, 2015). Therefore, in this case, high specific stiffness and strength, or ratios such as specific modulus (E/ρ), where ρ is the density, and specific strength (σ_f/ρ), respectively, are used to consider more than one criterion when selecting suitable materials for aerospace applications. Other applications allocate priorities to different material properties, for example, automotive passenger cars are very cost sensitive, and some applications place nontechnical properties above technical properties, eg, consumer goods.

Good knowledge about material property data is essential in order to stipulate conditions on the properties. When designers and engineers have decided on the important design criteria, the combination of parameters which best describes them may be derived as a material index. Performance indices have simplified the problem of rigorously comparing different materials for many common mechanical and thermal loading situations, and allowing technical properties (eg, strength) and nontechnical properties (eg, cost) to be simultaneously considered. Deriving material performance indices has been well described in the literature, with tables of commonly used indices published (Ashby and Cebon, 1995). Design factors can be classified simply into objectives and constraints. Objectives are aims or targets to be achieved by the designer such as reducing the mass or increasing the stiffness. The degree to which these objectives are achieved will be dictated by the constraints. It is these objectives and constraints that may be used to decide on which material indices need to be used as shown in Fig. 3.2 (Ashby et al., 2004).

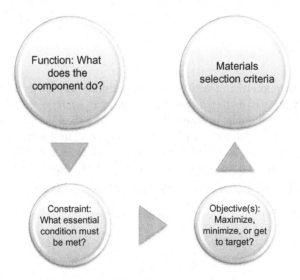

Figure 3.2 Design analysis and determining materials selection criteria.

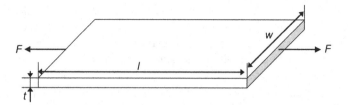

Figure 3.3 Element of sheet undergoing a uniaxial load.

The identification of suitable material indices (combining function, constraint, objective, solid mechanics, and materials science) is essential in the materials selection process. These indices might result from concerns about the stress conditions and/or the geometry of a particular component. As an example, Fig. 3.3 demonstrates an element of a sheet or plate subject to an in-plane uniaxial load. The objective in this case is to minimize the mass (m) of the sheet as defined in Eq. (3.1).

$$m = \rho.V = \rho.w.t.l \qquad (3.1)$$

where V is the volume, w is the width, t is the thickness and l is the length of the sheet, and ρ is the density of the material.

The sheet carries a tensile force or load F, and σ_f is the ultimate tensile strength of the material. The requirement that the material must satisfy is to sustain the load F as expressed by Eq. (3.2).

$$\frac{F}{w.t} \leq \sigma_f \qquad (3.2)$$

By combining Eqs. (3.1) and (3.2) and considering the sheet thickness t as a free variable, it is possible to describe the performance parameter by means of Eq. (3.3).

$$F.l.\frac{\rho}{\sigma_f} \leq m \qquad (3.3)$$

Eq. (3.3) is therefore the objective function for this specific engineering problem. The design objective is to minimize the mass m that is subjected to the condition expressed in Eq. (3.3) in order to ensure the ability of the sheet to support the load F. In other words, there is a need to maximize the performance, in this case minimize mass m, while at the same time achieving the functional and geometric requirements, in the assumption that both the load F and the geometric term l are given by the problem-specific constraints and requirements. This means obtaining the smallest value of $\frac{\rho}{\sigma_f}$ or the largest value of Eq. (3.4), which is defined as the performance index.

$$\operatorname*{Max}_{i \in M} \left(\frac{\sigma_f}{\rho}\right)_i \qquad (3.4)$$

where the subscript i indicates the ith material within a discrete set of alternative materials.

Types of criteria

Individual material properties	Indices made by combination of individual material properties	Indices made by combination of material properties and geometry	Nontechnical criteria (eg, Cost or availability for long-term)

Figure 3.4 Types of material selection criteria.

Table 3.1 Effect of performance index on the selected material: Importance of weighting

Material	E (elastic modulus)	ρ (density)	C (cost)	$M_1 = \frac{E}{\rho C}$
Objective	Max.	Min.	Min.	Max.
Material 1	12,000	5	50	48
Material 2	12,000	5	40	60
Material 3	12,000	4	50	60
Material 4	10,000	4	40	62.5

It is possible to create many different performance indices to address the interaction of material and function. Each index or criterion combines a defined set of properties (eg, tensile strength, elastic modulus, density, cost, etc.) and is used to evaluate candidate materials for a specific application. A typical application might make use of existing criteria and/or use newly developed criteria. Fig. 3.4 shows the different types of materials selection criteria. The choice of criteria is mostly dictated by the application. The situation becomes complicated when several criteria need to be investigated simultaneously.

It is worth mentioning that depending on the application, the material properties might have different importance (or weighting) in practice. For example, in biomedical applications such as prostheses, the cost should be considered as well as the performance of the material like nontoxicity and an elastic modulus to closely match with that of bone is of crucial importance. Therefore, when selecting materials higher weightings must be assigned to those criteria. It is therefore useful to see how performance indices can be affected by neglecting the importance of material properties. For demonstration purposes only of the formalized materials selection process, Table 3.1 shows the sensitivity of a simple performance index on its constituent properties for four fictitious materials. It can be seen that Material 3 has the best technical properties while the performance index, M_1, suggests that Material 4 is the best material because of the lower cost. It can be concluded that a strong

Table 3.2 Effect of performance index formula on material ranking

Material	Stress (A)	Yield strength (B)	B/A	B − A
Material 1	10	40	4	30
Material 2	15	40	2.67	25
Material 3	100	400	4	300
Material 4	20	400	20	380

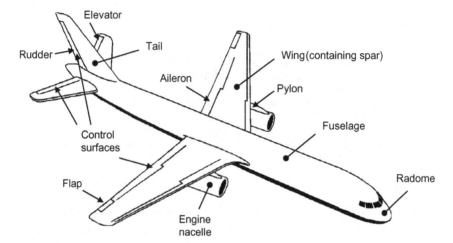

Figure 3.5 Schematic of aircraft main structures.

decision on the final material choice can be achieved through weighting the relationship between elastic modulus, density, and cost.

Furthermore, performance indices can be affected by the mathematical formulation used. Table 3.2 compares a simple ratio-based performance index with a subtraction-based index. It is clear that in the case of comparing stress and yield strength for the four different fictitious materials, the ratio of B/A cannot make a clear distinction between Materials 1 and 3, while in the last column the superiority of Material 3 is clear. The use of mechanics of materials analysis will also determine particular mathematical formulations for performance indices. This issue has been further discussed in conjunction with normalization techniques in materials selection (Jahan and Edwards, 2015).

Therefore, when selecting materials where numerous choices with similar performance exist and a lot of different criteria influence the selection, utilizing a more accurate performance index is essential for identifying the most suitable material.

For a more realistic example, if the aircraft structure application is used again, in this case an aircraft fuselage (Fig. 3.5), the following materials selection process has been considered to help further demonstrate the underlying principles involved. The example described is not meant to provide a definitive answer, which will need

Table 3.3 Data for materials used in aircraft fuselage

Material	σ_y (MPa)	E (GPa)	K_{IC} (MPa m$^{1/2}$)	ρ (Mg/m^3)	Cost ($/kg)
Stainless steel	990	210	98	7.80	7
Titanium alloy	855	110	62	4.45	50
Aluminum alloy	450	70	35	2.75	2

Table 3.4 Overall ratings for the materials used in aircraft fuselage

Material	$M_1\ \sigma_y/\rho$ MPa/(Mg/m^3)		$M_2\ E^{1/3}/\rho$ (GPa)$^{1/3}$/ (Mg/m^3)		$M_3\ (K_{IC}/\sigma_y)^2$ m × 10^{-3}		C (cost) $/kg		Overall rating [4M_1 + 2M_2 + 3M_3 + (1 − C)]/10
	ABS	REL	ABS	REL	ABS	REL	ABS	REL	
Stainless steel	126.92	0.66	0.76	0.51	9.80	1.00	7	0.14	0.75
Titanium alloy	192.14	1.00	1.08	0.72	5.26	0.54	50	1.00	0.71
Aluminum alloy	163.64	0.85	1.50	1.00	6.05	0.62	2	0.04	0.82

ABS, absolute; REL, relative (normalized value).

much detailed investigation in practice, but merely to illustrate how the process highlights issues, some intuitive, that typically need to be addressed as part of the decision-making process of designers. For the design of an aircraft fuselage, the following material properties are important: yield strength (σ_y), elastic modulus (E), fracture toughness (K_{IC}), and anything to do with aircraft, density (ρ). Materials to be considered are: Aluminum alloy, Titanium alloy, and Stainless steel, of which there are many grades but for simplicity, the average properties are used only. The pertinent data for the proposed materials to be used in the aircraft fuselage are shown in Table 3.3. It should be noted that the cost has been approximated from data published in trade journals at the time of writing, which unlike the materials properties will vary with time, but serve adequately for the purpose of comparison.

The relevant performance indices, M_1, M_2, and M_3 combining the material properties from Table 3.3 are shown along with cost again for convenience in Table 3.4. The performance indices have different units and to allow direct comparisons to be made, each performance index has been normalized by expressing its value relative to the largest value. The last column combines the weighted relative performance indices as an overall rating for each material. The overall rating formula has allocated different weightings to the relative performance indices and when placed in a spreadsheet can easily be changed by the designer to check the sensitivity of the results. For a different application the overall rating formula would have a different set of relative performance indices and allocated weightings.

Clearly, the materials selected for the application in this case are well established and therefore the outcome is not a surprise, justifying the widespread use of aluminum alloys in commercial aircraft structures. A more detailed investigation would also include fatigue properties and other materials such as composites. The situation will be different for military aircraft structures of course, with different weightings for the performance indices and use of other performance indices that include other important material properties such as thermal properties. The mechanical performance will also have a much higher priority than cost, and the weightings altered accordingly in the overall rating formula. The next stage would be to repeat the process to refine the selection further, but this time the list of materials will be replaced with various grades of aluminum alloy, allowing a more specific aluminum alloy to be selected. The whole process can be repeated again for the rest of the aircraft structure, recognizing that the property requirements will vary for the different parts of the aircraft. This may result in different parts of the aircraft structure being made of different materials in order to optimize the performance of the airframe, for example, in a modern civil aircraft, the fuselage is now made from aluminum alloy and wings are made from carbon fiber composite, instead of all the airframe being made from aluminum alloy in older aircraft. The use of any new material though has to be balanced against the risks, particularly for use in aircraft where safety is paramount. The example does clearly illustrate the complex nature of the decision-making involved when considering multiple materials selection criteria in practice, for which computer-based implementations of the processes described can be helpful.

3.4   Brief review of current formalized screening methods

There are many materials available and a lot of information to consider on these materials. Therefore, it is necessary to screen materials. When selecting materials, formalized screening methods help to narrow down the choice of materials to a more manageable number for subsequent detailed evaluation. A variety of procedures, as shown in Fig. 3.6, have been developed to solve this issue (Jahan et al., 2010; Dieter, 1997; Ashby et al., 2004). These methods are reviewed briefly below. The chart method and computer-based materials selection systems are described later in Sections 3.5 and 3.6.

3.4.1   Cost per unit property method

Since cost is so important in selecting materials, it is logical to consider it at the start of the materials selection process. Generally, a target cost is set to remove the materials that are very expensive. Therefore, the final choice of material becomes a trade-off between cost and performance. This method uses a special type of performance index and it can be the most useful factor when it is related to a critical

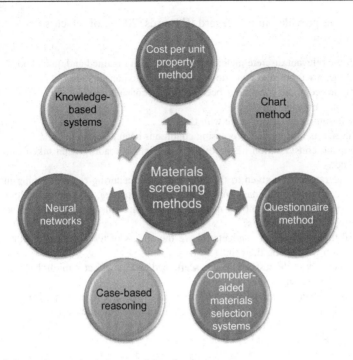

Figure 3.6 Screening methods in materials selection.

material property that controls the performance of the design. The cost per unit property method is appropriate for initial screening in applications where one property stands out as the most critical service requirement (Farag, 1979). In this case, it is possible to estimate the cost of satisfying the most critical requirement for various materials. Cost/unit tensile strength ($$/MPa) is usually one of the most regularly used criteria and materials with lower cost/unit strength in this case are preferable. However, the main limitation of this method is that it considers only one property, ignoring other properties, which have to be considered separately.

It is also very important to highlight that there are many ways to compute costs. Total life cycle cost is the most appropriate cost to be considered. This cost consists of the initial material costs, manufacturing costs, operation costs, and maintenance costs.

3.4.2 Questionnaire/guideline method

Most current materials selection methods are based on numerical data that become more detailed as the designer progresses through the stages of the design process. Experts emphasize the need for materials selection tools to support designers more at the early stages of the design process in order to develop either feasible or more economic design solutions (Deng and Edwards, 2007; Van Kesteren et al., 2007a). Although the questionnaire or guideline method cannot offer any ranking, it improves the likelihood of achieving an optimal design solution. A set of structured questions were provided in this regard (Edwards, 2005), of which some of them are mentioned below:

- Have all the relevant discrete material properties been obtained and understood?
- Have all the environmental conditions been taken into consideration?
- Have all the economic constraints been taken into consideration?
- Are there any transient effects?
- Will the design conditions change with time?
- Have the effects of materials processing conditions been considered?
- Has adequate consideration of the effects of dynamic loading been taken into account, eg, fatigue?
- Has consideration been given to the effects of manufacturing (eg, machining introducing stress concentrations or damage) on the change of properties, both short and long term?
- Will the properties of the chosen material change when the component is in service as a result of operating conditions?
- If special testing has been conducted, have the effects of test conditions been taken into consideration when using the results?
- Has the accuracy of the test data been ascertained and the effect it might have on the quality of design solution?
- Has a risk assessment been conducted for the effects of inaccurate or missing information?
- Has the effect of any postprocessing such as surface treatment been taken into account on the final material properties?
- Has adequate consideration been given to alternative materials?
- Has the overall life cycle been adequately considered, eg, recycling principles?
- Has the complete design situation been considered where the component is an integral part of a product and decisions made in isolation may lead to a suboptimal product design?
- Has consideration been given to the interfaces with other components such as wear, joining techniques, corrosion, etc.?
- Are there any chemical, toxicological, radiation, or microorganism effects that need to be considered?
- Have the effects of quantities and rate or production of components been adequately considered?
- Has future raw material availability been considered?
- Has the introduction of any new materials and technologies been considered, balancing their risks and lack of familiarity against potential benefits?

For user-interaction aspect of materials, Van Kesteren et al. (2007b) proposed a questionnaire tool. The tool is composed of a list of questions for the different phases in user–product interaction and a checklist of sensorial properties. It assists clients in clearly specifying requirements that relate to user-interaction and form consensus between them and the product designer about these aspects in early stages of materials selection. The tools are only able to translate a low percentage of user-interaction aspects into sensorial properties.

3.4.3 Knowledge-based systems

Knowledge-based systems (KBS), or expert systems, are computer-based systems utilizing artificial intelligence methods and techniques. KBS imitate human problem-solving and reference a database of knowledge on a particular subject. The information on engineering materials is presented in two categories: data and knowledge. However, the distinction between a database and a knowledge base is still unclear (Trethewey et al., 1998). "Data" are defined as the result of measurements that can be presented in numbers, whereas "knowledge" represents the connections between items of data and is mostly expressed in plain language. According to these categories, there are two possibilities to maintain the process of materials selection, via material databases or KBS. Computerized databases are the best form of presenting the former because of providing easy access to the materials' data. The latter comprises expert knowledge capable of assisting the user in an interactive way to solve different problems and queries (Ermolaeva et al., 2002). The KBS works in a fully interactive mode (Sapuan et al., 2002) and provides impartial recommendations, and is able to search large databases for optimum solutions (Farag, 2002). Another important benefit of KBS is their ability to capture valuable expertise and make it available to a wide range of users (Farag, 2002). However, knowledge elicitation is a difficult process and it is not easy to maintain this system. Furthermore, these systems are not suitable tools for ranking.

3.4.4 Neural network

A neural network applied to materials selection can be useful in helping to select the best alternative material from a database but used on its own it is not sufficient because in some cases it may not be able to provide a unique solution (Goel and Chen, 1996).

3.4.5 Case-based reasoning

Case-based reasoning (CBR) is a useful technique for searching databases with information on different technical solutions, including materials use in product design, already applied in practice and the results of past failures, which can be easily updated. CBR systems can learn from experience by acquiring new knowledge from different cases thus making maintenance easier. Critics of CBR though debate that it is an approach that accepts anecdotal evidence without the backing of statistically relevant data. Using implicit generalization, there is no guarantee that the solution will be correct.

3.5 Materials' property charts (the "Ashby" approach)

The materials selection method developed by Ashby (1992) concentrates on the data modeling aspect of a mechanical design problem by presenting the materials' data in a graphical chart format. Ashby's materials selection charts (Ashby, 1992, 1997b) are helpful for the initial screening of materials, and aimed at comparing different materials on the basis of more than one material property. Materials

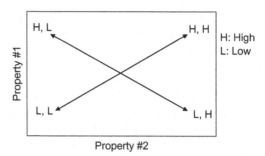

Figure 3.7 Principles of the chart method.

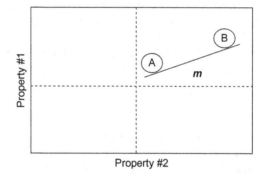

Figure 3.8 Identifying competing materials in the chart method.

selection charts are very useful in showing trade-offs between two material properties as shown in Fig. 3.7 (Ashby, 2005).

This method presents a lot of different materials, allowing easy visualization of their properties as well as demonstrating the balance of different properties (eg, strength vs cost) and it is therefore ideal as a first "rough cut" selection. The charts can therefore be used to initially identify the best classes of materials, and then used to look in more detail within these classes in order to evaluate competing candidate materials as shown in Fig. 3.8.

The number of materials that can be viewed in this way is limited and the process can quickly become very unwieldy as the number of constraints increase. To address these problems, the methodology has been successfully implemented as a commercial computer software tool. The Cambridge Engineering Selector (CES) is a powerful selection and analysis tool that is based on Ashby's materials selection methodology.

In the field of mechanical engineering design the materials selection charts are a simple and quick way of assessing whether a material is suitable for a particular application (Holloway, 1998). On these charts (see Fig. 3.9 as an example), materials of each class (eg, metals, ceramics, polymers, etc.) form "clusters" or "ballons." It is clear to see the material families and how the properties, Young's modulus and

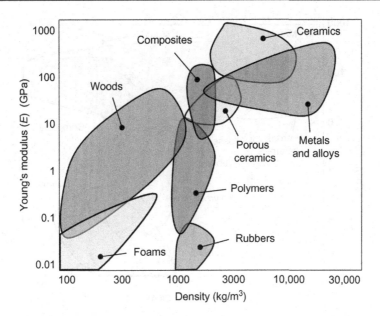

Figure 3.9 Example materials selection chart: Young's modulus versus density.

density in this case, are similar in each group. It can also be seen that metals are the heaviest materials, foams are the lightest materials, and ceramics are the stiffest materials. Presenting the information in this graphical format allows many materials to be easily compared at a glance, allowing competing materials to be quickly identified. For instance, if a light and stiff material is required, materials need to be chosen near the top left corner of the chart—therefore composites look good. Figs. 3.10 and 3.11 show other examples of material property selection charts.

It should be noted that most material selection charts use logarithmic scales because material properties often cover large ranges of values and it is essential to show the full range of properties on one chart for ease of comparison. A good way to use the charts is to eliminate materials that will definitely not be good enough, rather than to try and identify the single best material too soon in the design process. The use of logarithmic scales also allows more information to be shown on the chart that can be used for design purposes. All designs have primary constraints that must not be exceeded and these appear as straight vertical or horizontal lines on the chart. For a light stiff component for example, there will be critical limits on the Young's modulus and density, as shown in Fig. 3.12, with only materials above the horizontal line and to the left of the vertical line being acceptable. From the resulting subset of materials satisfying the primary constraints, it is then necessary to find the materials that maximize the performance of the component, using a performance index as previously described in Section 3.3. These performance indices (the ratio of vertical and horizontal axis variables equated to a constant) appear as groups of parallel straight lines with the same slope for each constant with a lower limit being the intersection of the horizontal and vertical constraint lines.

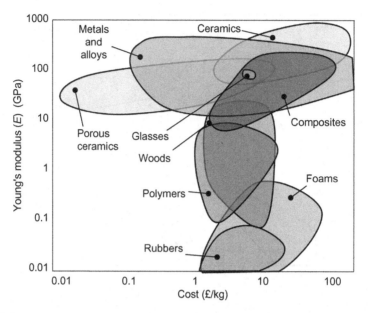

Figure 3.10 Example materials selection chart: strength versus cost.

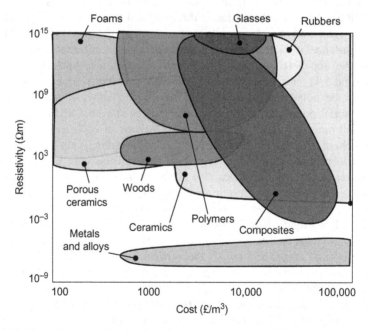

Figure 3.11 Example materials selection chart: strength versus maximum service temperature.

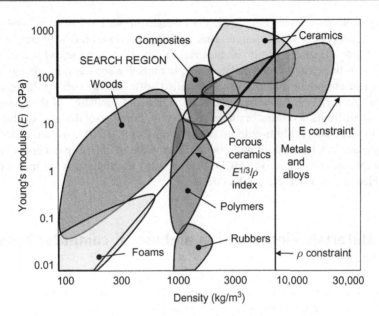

Figure 3.12 Example use of primary constraints and performance index for light stiff component.

For example, light stiff components: E/ρ = constant for a slope of 1, $E^{1/2}/\rho$ = constant for a slope of 2, and $E^{1/3}/\rho$ = constant for a slope of 3, for maximizing vibration frequencies for ties, beams, and plates, respectively. All the materials that lie along each sloping line will perform equally as well, with all materials appearing above each sloping line being acceptable and all materials appearing below each sloping line not being acceptable. For the light stiff sheet type component from the aircraft structure example in Section 3.3 (airframe skins), there will be a critical limit on $E^{1/3}/\rho$ as shown in Fig. 3.12. Used in conjunction with the primary constraint lines, minimum Young's modulus E and maximum density ρ along with the sloping index line $E^{1/3}/\rho$ = constant defines a portion of the chart (top left hand corner) with a smaller subset of acceptable materials, which is the material search region for design purposes. If the same aircraft structure example is used, the process has to then be repeated for any other relevant constraints such as yield strength σ_y and fracture toughness K_{IC} and performance indices σ_y/ρ and $(K_{IC}/\sigma_y)^2$ on σ_y versus ρ and K_{IC} versus σ_y charts, respectively. This over constrained example is a typical materials selection problem and a pragmatic way forward is to rank the constraints in order of importance, investigating each constraint in turn gradually reducing the subset of suitable materials in the process, or apply weightings as shown in Section 3.3 to calculate an overall rating.

The benefit of using this approach is that it is systematic and impartial in its focus on the product design objectives. The selection of materials derived from the charts is normally left quite broad to initially keep options open and then narrowed down as the design develops. Also, as the number of possible materials reduces, the information about the materials gradually changes in the level of detail and precision, from concept to detail design. This further exacerbates the difficulty in being able to decide the most appropriate material to use. The chart method is easy when the design of the component specifies a simple objective such as minimizing weight and a single constraint, for instance a specified stiffness, strength, or thermal conductivity (Ashby, 1997a). Perhaps the most significant limitation of this method is that the chart limits decisions in materials selection to only solving two or three criteria simultaneously. Multi-criteria decision-making (MCDM) is an established technique that can be used for addressing the problem of choosing materials involving multiple criteria. Therefore, MCDM can be used to supplement the use of charts when selecting materials for product design.

3.6   Materials identification and use of computer-based tools

Traditional textbooks alone cannot encompass the wealth of information now accessible to the designers and materials science and engineering students, and therefore the natural response is to turn to the Internet for more information (Silva and Infante, 2015). Being able to easily compare material properties and then select the most appropriate materials helps designers to enhance the performance of products. Therefore, it is important to always consider all materials and retain a wide range of materials at the ranking stage. What is important in materials selection are data for a lot of material properties for a lot of materials. This information normally comes as tables of data and it can be a very time consuming and difficult process to sort through them. With a very large number of materials there is a self-evident need for information-management systems (Ashby et al., 2004). Although some of the materials databases can also be used as materials selection systems, they are essentially developed for data storage searching only. Electronic materials databases and data search/retrieval computer software can help designers in this regard. When accessed "Online" via the Internet, these databases allow designers to keep up to date and be made aware of the main advances in materials. It also enables designers to access information about a greater range of different materials, all of which can be very useful for designing new competitive products. However, there is always some mistrust about the information resources taken from Internet websites that have not been peer reviewed as is normally expected from scientific literature (Silva and Infante, 2015).

Using new materials is an important way of innovating in product design. New materials may either be completely new to the market or existing materials just new to a particular designer or organization. This is the advantage of being able to consider a large number of possible materials at the start of the design process (concept stage). It allows the possibility of being able to think about materials not used before and make comparisons with materials already being used. However, it brings with it risks associated with a lack of familiarity of the new materials. The problem is particularly acute when considering alternative materials in different materials classes, eg, metals to polymers. Often new materials create the need to understand new properties (and processes). In the case of using polymers, thermal properties and resistance to creep are important and these may not have been such an issue if previously metals were used. The different processing routes also cause concerns as well as opportunities. Unfortunately, there is limited effort being spent in supporting the identification of new materials in the early stages of the design process (Deng and Edwards, 2007).

In the case of technical and economical requirements data, the upper and lower limits of material properties for instance, can be entered into material databases such as *MatWeb* and *CES* for filtering a large number of material options that meet all the design requirements. A critical analysis of the digital tools available for materials selection has been carried out by Ramalhete et al. (2010). This was done by answering a few important questions. What kind of digital tools exist? How do they work? What properties determine the selection? What kind of information results from the selection? This kind of information is very helpful to product designers, materials engineers, and associated practitioners performing materials identification and selection using computer-based tools for product design.

3.7 Summary and conclusions

The different materials selection methods, including KBS, Questionnaire/guideline method, computer-aided materials selection systems, and neural-networks are used for material screening, but these techniques do not provide any ranking order. The cost per unit property method only considers one material property as the most critical and ignores other material properties. The chart (Ashby) method, which is the most popular screening method, limits the decisions in materials selection to only two or three criteria. Therefore, the traditional screening approaches cannot guarantee the selection of the best material. MCDM can be used to supplement the use of material screening methods, particularly for the chart method, when selecting materials for new product design.

Review questions

1. Why is it useful to screen materials in the early stages of the design process?
2. Why is it difficult to screen materials based on qualitative information, and what method is appropriate to help guide materials selection under such circumstances?
3. Explain why the chart method of materials selection is popular even though there are a variety of other methods available for screening materials?
4. What are the advantages of using computer-based or digital tools when selecting materials, whatever the stage of the design process?
5. Why does having to satisfy more than one performance index lead to the need to compromise in materials selection?

6. Why is it necessary to have good knowledge and understanding of materials' properties when specifying materials selection criteria?

7. How can weighting and formulating increase the precision of performance indices?

8. Why is it essential to consider the relationship between materials, processing and shape when designing components?

9. Why is it important for product designers to use and keep up to date with developments in formalized methods of selecting materials, particularly computer-based methods, even if they are experienced practitioners?

References

Ashby, M., Shercliff, H., Cebon, D., 2013. Materials: Engineering, Science, Processing and Design. Butterworth-Heinemann, Oxford.

Ashby, M.F., 1992. Material Selection in Mechanical Design. Pergamon Press, Cambridge, UK.

Ashby, M.F., 1997a. Materials selection: multiple constraints and compound objectives. ASTM Spec. Tech. Publ. 1311, 45–62.

Ashby, M.F., 1997b. Performance Indices. *ASM International*, Materials Park, OH.

Ashby, M.F., 2005. Materials Selection in Mechanical Design. Butterworth-Heinemann, Oxford.

Ashby, M.F., Cebon, D., 1995. A Compilation of Material Indices. Granta Design Ltd., Cambridge, MA.

Ashby, M.F., Johnson, K., 2013. Materials and Design: The Art and Science of Material Selection in Product Design. Butterworth-Heinemann, Oxford.

Ashby, M.F., Brechet, Y.J.M., Cebon, D., Salvo, L., 2004. Selection strategies for materials and processes. Mater. Des. 25, 51–67.

Crilly, N., Moultrie, J., Clarkson, P.J., 2004. Seeing things: consumer response to the visual domain in product design. Des. Stud. 25, 547–577.

Deng, Y.M., Edwards, K.L., 2007. The role of materials identification and selection in engineering design. Mater. Des. 28, 131–139.

Dieter, G.E., 1997. Overview of the materials selection process. ASM Metals Handbook, Materials Selection and Design, ASM International.

Edwards, K.L., 2005. Selecting materials for optimum use in engineering components. Mater. Des. 26, 469–473.

Edwards, K.L., Deng, Y.M., 2007. Supporting design decision-making when applying materials in combination. Mater. Des. 28, 1288–1297.

Ermolaeva, N.S., Kaveline, K.G., Spoormaker, J.L., 2002. Materials selection combined with optimal structural design: concept and some results. Mater. Des. 23, 459–470.

Farag, M.M., 1979. Materials and Process Selection in Engineering. Elsevier Science & Technology, London.

Farag, M.M., 2002. Quantitative methods of materials selection. In: Kutz, M. (Ed.), Handbook of Materials Selection London, John Wiley & Sons.

Goel, V., Chen, J., 1996. Application of expert network for material selection in engineering design. Comput. Ind. 30, 87–101.

Holloway, L., 1998. Materials selection for optimal environmental impact in mechanical design. Mater. Des. 19, 133–143.

Jahan, A., Edwards, K.L., 2015. A state-of-the-art survey on the influence of normalization techniques in ranking: improving the materials selection process in engineering design. Mater. Des. 65, 335–342.

Jahan, A., Ismail, M.Y., Sapuan, S.M., Mustapha, F., 2010. Material screening and choosing methods—a review. Mater. Des. 31, 696–705.

Karana, E., Hekkert, P., Kandachar, P., 2008. Material considerations in product design: a survey on crucial material aspects used by product designers. Mater. Des. 29, 1081–1089.

Karana, E., Hekkert, P., Kandachar, P., 2009. Meanings of materials through sensorial properties and manufacturing processes. Mater. Des. 30, 2778–2784.

Lovatt, A.M., Shercliff, H.R., Withers, P.J., 2000. Material Selection and Processing. Technology Enhancement Programme, London.

Ramalhete, P.S., Senos, A.M.R., Aguiar, C., 2010. Digital tools for material selection in product design. Mater. Des. 31, 2275–2287.

Sandstrom, R., 1994. TALAT Lecture 1502 Criteria in Material Selection. European Aluminum Association, Stockholm.

Sapuan, S.M., Jacob, M.S.D., Mustapha, F., Ismail, N., 2002. A prototype knowledge-based system for material selection of ceramic matrix composites of automotive engine components. Mater. Des. 23, 701–708.

Silva, A., Infante, V., 2015. The role of digital libraries in teaching materials science and engineering. Handbook of Research on Recent Developments in Materials Science and Corrosion Engineering Education, p. 190.

Trethewey, K.R., Wood, R.J.K., Puget, Y., Roberge, P.R., 1998. Development of a knowledge-based system for material management. Mater. Des. 19, 39–56.

Van Kesteren, I., Stappers, P.J., De Bruijn, S., 2007a. Materials in products selection: tools for including user-interaction in materials selection. Int. J. Des. 1, 41–55.

Van Kesteren, I.E.H., Stappers, P.J., De Bruijn, J.C.M., 2007b. Defining user-interaction aspects for materials selection: three tools. In: The Web Proceedings of Nordic Design Research.

Multi-criteria decision-making for materials selection

4

Learning Aims

The overall aim of this chapter is to introduce the place of operations research and multi-criteria decision-making (MCDM) in materials engineering and design. After carefully studying this chapter you should be able to understand:

- the characteristics of multi-attribute decision-making and multi-objective decision-making methods used in decision-making problems
- the role of graphical- and mathematical-based methods in ranking of materials
- the justification for using multi-criteria analysis in material design and selection
- the importance of applying MCDM in combination with design tools and techniques.

4.1 Introduction to multi-criterion decision-making

The rapid advancement in technology is linked closely to the ability to engineer materials that meet the increasingly ambitious requirements of new products (McDowell et al., 2010a). The selection of the most suitable material for the design of a component is a difficult process that demands the management of a large amount of information about material properties (and processes) and there are often several solutions for a particular application (Chiner, 1988). After significantly reducing the number of possible materials via one or more of the initial screening methods described in Chapter 3, ranking methods can then be used to further narrow down the field of possible materials to a few likely candidates. multi-criterion decision-making (MCDM) methods and optimization approaches are useful for ranking purposes; especially when complex multi-criteria need to be considered simultaneously.

4.2 MCDM as a subdiscipline of operations research

"Operations research" is a relatively young discipline, being organized as a separate professional field of study only since the end of World War II. It is a discipline that deals with the application of advanced analytical methods to help make better decisions and is often considered to be a subfield of mathematics. Operations research arrives at optimal or near-optimal solutions in complex decision-making problems; it gives decision-makers the ability either to choose the

Multi-criteria Decision Analysis. DOI: http://dx.doi.org/10.1016/B978-0-08-100536-1.00004-7

"best" outcome or to enhance the likelihood of a given set of desired outcomes. Because of its emphasis on human−technology interaction and because of its focus on practical applications, operations research overlaps with other disciplines, particularly operations management and engineering science. Decision theory and optimization are closely linked in that decision-making and optimization contain exactly the same elements, and all decisions involve some amount of optimization (Hazelrigg, 2003).

MCDM has become one of the most important and fastest growing subfields of operations research/management science. MCDM means the process of determining the best feasible solution according to established criteria and problems that are common occurrences in everyday life. Practical problems are often characterized by several noncommensurable and conflicting (competing) criteria, and there may be no solution satisfying all criteria simultaneously. Hwang and Yoon (1981) suggest that MCDM problems can be classified into two categories:

- multi-objective decision-making (MODM)
- multi-attribute decision-making (MADM)

There are several methods in each of the above categories. Each technique has its own characteristics and the methods can combine with each other or with "fuzzy" methods. MODM methods have decision variable values that are determined in a continuous or integer domain with either an infinitive or a large number of alternative choices. MODM can usually be fitted in programming/designing facet, which is to achieve the optimal goals by considering the various interactions within the given constraints. Almost all multi-objective optimization (MOO) problems can be mathematically represented as:

$$\text{Min}\{f_1(x), f_2(x), \ldots, f_k(x)\}$$
$$\text{subject to} \quad x \in S$$

where k (≥ 2) is the number of (conflicting) objective functions, $x = decision\ vector$ (of $n\ design\ variables\ x_i$), $S \subset R^n = feasible\ region$ formed by $constraint\ functions$, and S consists of linear, nonlinear (equality and inequality), and lower and upper bounds for the variables. In the objectives, maximize $f_i(x) = $ minimize $-f_i(x)$.

The x vectors are said to be Pareto-optimal for a multi-objective problem if all other vectors have a higher value for at least one of the objective functions, or have the same value for all the objective functions. The image of all the efficient solutions is called a Pareto-front or Pareto curve or surface that indicates the nature of the trade-off between the different objective functions. An example of a Pareto curve is shown in Fig. 4.1, where all the points on the bold lines define the Pareto-front. These points are called noninferior or nondominated points. Points P1 and P5 are called weak Pareto-optima and points P2, P3, and P4 are strict Pareto-optima. Finding Pareto frontiers is particularly useful in engineering. By restricting attention to a set of choices, the potentially optimal solutions that are Pareto-efficient, a designer can make trade-offs within this set, rather than considering the full range of alternatives.

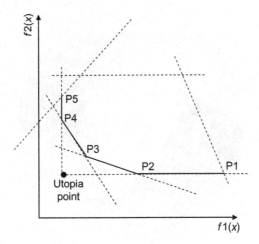

Figure 4.1 Pareto curve and concept for weak and strict Pareto-optima.

Table 4.1 A typical multi-attribute decision-making problem

	w_1 C_1	w_2 C_2	$\ldots$	w_n C_n
A_1	x_{11}	x_{12}	$\ldots$	x_{1n}
A_2	x_{21}	x_{22}	$\ldots$	x_{2n}
A_3	x_{31}	x_{32}	$\ldots$	x_{3n}
$\vdots$	$\vdots$	$\vdots$		$\vdots$
A_m	x_{m1}	x_{m2}	$\ldots$	x_{mn}

Alternatively, MADM methods are used for making preference decisions over the available alternatives, which are characterized by multiple (usually conflicting) attributes. MADM are generally discrete, with a limited number of prespecified alternatives. An MADM problem with finite possibilities can be expressed in a matrix format. It includes possible alternatives (materials) A_i ($i = 1, \ldots, m$), which decision-makers have to choose, criteria (material properties) C_j ($j = 1, \ldots, n$), relative importance of criteria (or weightings) w_j, and elements of x_{ij}, which is the rating of alternative i with respect to criterion j as shown in Table 4.1.

4.3 Graphical-based ranking methods used in materials selection

The use of graphical-based ranking methods is best described using an example of a materials selection problem. Suppose the problem under consideration is selecting bearing materials, and to rank a number of materials with different properties. A bearing in general is a machine element that transmits loads or reaction forces

Figure 4.2 Example of a plain journal bearing.

from a moving component to a bearing support. Journal or cylindrical bearings are used when the transmitted load is essentially perpendicular to the axis of a rotating shaft or journal. In addition to carrying these radial loads, the bearing material is also subjected to sliding movement of the shaft. The main problems in selecting suitable bearing materials is ensuring sufficient strength to sustain the applied load without deformation and sufficient hardness and wear resistance for an adequate service life but avoiding wear of the journal. An example of a plain journal bearing or bush is shown in Fig. 4.2. The bearing can be manufactured by either casting, or drawing and machining, or powder metallurgy process, and is normally assembled as a press-fit into the bearing support (bracket, housing, etc.) and the bore finished by boring and reaming to match the diameter of the journal.

Looking at the particular plain journal bearing example in more detail, a preferred bearing material likely needs to meet the following requirements as described in an extract from a design case study by Farag (1997):

- The compressive strength of the bearing material at the operating temperature (120°C in this case) must be sufficient to support the load acting on the bearing. From design calculations, the minimum allowable room-temperature compressive strength is 20 MPa.
- Fatigue strength is important under conditions of fluctuating load with the minimum allowable of 20 MPa.
- Conformability of the bearing material allows it to change its shape to compensate for slight deflections and misalignments. A low Young's modulus is desirable with the maximum allowable of 100 GPa.
- Embed ability is the ability to embed grit or similar foreign particles to prevent them from scoring the journal. A low hardness is desirable and the maximum allowable is 100 BHN.
- Wear resistance is a lower-limit property, which can be rated along a scale as excellent (5), very good (4), good (3), fair (2), or poor (1).
- Corrosion resistance is a lower-limit property, which can be rated along a scale as excellent (5), very good (4), good (3), fair (2), or poor (1).

- Thermal conductivity is also important and the minimum required conductivity is 20 W/m K.
- Low cost is desirable and includes the cost of bearing material and manufacture.

This is a typical set of design requirements for a plain bearing material. Many metal alloy systems have been specially developed to accommodate the conflicting requirements that must be satisfied by plain bearing materials. These requirements are met by the use of soft metal alloys with embedded hard particles. The bearing materials can be classified as:

1. white metals, which are either tin base or lead base
2. copper-base bearing alloys of a wide range of strengths and hardness
3. aluminum-base bearing alloys, which have good thermal conductivity.

A list of the candidate bearing materials, the beneficial and nonbeneficial criteria, and the values of the attributes are shown in Table 4.2.

The data for the bearing material properties and selection criteria can now be used to demonstrate the use of graphical charts for materials selection. Consideration might be given to using a Column (bar), Trend (line), or Spider (network) chart to evaluate the performance of materials graphically. However, the scale or units of the data must first be eliminated before demonstrating them on any graph. Therefore, the linear max−min normalization method can be used to eliminate the dimensions of the selection criteria for both cost and benefit (Eqs. (4.1) and (4.2), respectively).

$$n_{ij} = \frac{x_j^{max} - x_{ij}}{x_j^{max} - x_j^{min}} \tag{4.1}$$

$$n_{ij} = \frac{x_{ij} - x_j^{min}}{x_j^{max} - x_j^{min}} \tag{4.2}$$

Figs. 4.3, 4.4, and 4.5, show a bar chart, a connecting lines graph, and a spider diagram, respectively, for a few of the materials shown in Table 4.2. A bar chart displays data in separate columns and in Fig. 4.3 multiple bar charts have been used to compare eight criteria for five materials. A line graph as shown in Fig. 4.4 can compare multiple data sets easily. In a spider diagram as shown in Fig. 4.5, the values of different material attributes are plotted on individual axes that are finally joined to form a network or "spider web." Charts are a reasonably quick and simple way of visually presenting information that can be readily assimilated. It is easy to see which material for which criterion has the best or worst performance. However, when there are a large number of candidate materials and selection criteria to compare together (20 materials and 8 selection criteria in the case of the bearing material example), the charts become cluttered.

It is clear that the graphical methods described are highly visual but it can be difficult to identify the best material using them. This is because it is difficult to rank the

Table 4.2 **Properties of some bearing materials (Farag, 1997)**

Design objectives		Max.	Max.	Min.	Max.	Max.	Max.	Min.	Min.
No.	Materials	Yield strength (YS)	Fatigue strength (FS)	Hardness (H)	Corrosion resistance (CR)	Wear resistance (WR)	Thermal conductivity (TC)	Young's modulus (YM)	Relative cost (C)
1	White metals ASTM B 23 (tin base)—Grade 1	30	27	17	5	2	50.2	51	7.3
2	White metals ASTM B 23 (tin base)—Grade 2	42.7	34	25	5	2	50.2	53	7.3
3	White metals ASTM B 23 (tin base)—Grade 3	46.2	37	37	5	2	50.2	53	7.3
4	White metals ASTM B 23 (tin base)—Grade 4	38.9	31	25	5	2	50.2	53	7.3
5	White metals ASTM B 23 (tin base)—Grade 5	35.4	28	23	5	2	50.2	53	7.5
6	White metals ASTM 23 (lead base)—Grade 6	26.6	22	21	4	3	23.8	29.4	1.3
7	White metals ASTM 23 (lead base)—Grade 7	24.9	28	23	4	3	23.8	29.4	1.2
8	White metals ASTM 23 (lead base)—Grade 8	23.8	27	20	4	3	23.9	29.4	1.1
9	White metals ASTM 23 (lead base)—Grade 10	23.8	27	18	4	3	23.9	29.4	1
10	White metals ASTM 23 (lead base)—Grade 11	21.4	22	15	4	3	23.9	29.4	1

11	White metals ASTM 23 (lead base)—Grade 15	28	30	21	4	3	23.9	29.4	1
12	Copper-base alloys SAE (copper-lead)—Grade 48	40	45	28	3	5	41.8	75	1.5
13	Copper-base alloys SAE (copper-lead)—Grade 49	45	50	35	3	5	41.8	75	1.5
14	Copper-base alloys SAE (copper-lead)—Grade 480	38	42	26	3	5	41.8	75	1.5
15	Copper-base alloys ASTM B22 (bronze)—Grade A	168	120	100	2	5	41.8	95	1.8
16	Copper-base alloys ASTM B22 (bronze)—Grade B	126	100	100	2	5	41.8	95	1.8
17	Copper-base alloys ASTM B22 (bronze)—Grade C	119	91	65	2	3	42	77	1.6
18	Aluminum-base alloys—Grade 770	173	150	70	3	2	167	73	1.5
19	Aluminum-base alloys—Grade 780	158	135	68	3	2	167	73	1.5
20	Aluminum-base alloys—Grade MB7	193	170	73	3	2	167	74	1.5

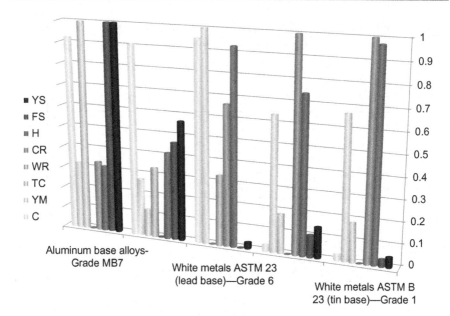

Figure 4.3 Bar chart for comparison of bearing materials.

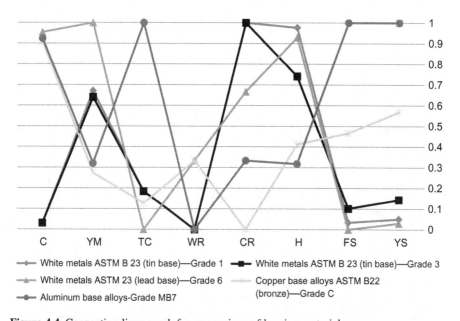

Figure 4.4 Connecting lines graph for comparison of bearing materials.

importance of the different selection criteria for each material, which gets worse with the number of materials and selection criteria involved. The graphical methods described are therefore limited to problems with ranking only a small number of candidate materials and selection criteria. Despite the appeal therefore of visual presentation

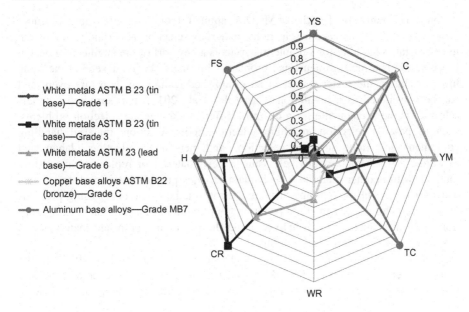

Figure 4.5 Spider diagram for comparison of bearing materials.

of graphical-based methods for ranking of materials in detail design, they are ultimately impractical for all materials selection problems because of an inability to effectively communicate large amounts of data and complicated relationships between alternative materials and selection criteria. It is therefore self-evident, except for simple materials selection problems, or identifying extremes or trends in data sets, or investigating small subsets of materials and/or selection criteria, to apply more advanced analytical methods for ranking of materials in materials selection problems. In this respect, graphical methods can supplement analytical methods for ranking materials because products in general break down into several components, with each requiring different levels of analysis in order to select materials. There is a significant benefit though of graphical methods in materials selection, which are most useful in the early stages of the design process when material information is limited in precision and detail, even though the number of materials is large, hence the popularity of the use of "Ashby" materials selection charts, described in Chapter 3.

4.4   Justification for applying MCDM in materials selection

Most real-life decision-making problems have several conflicting criteria and objectives to be considered simultaneously. For example, the compromises required to strike a balance between the performance and cost of a motor car, or between health and the pleasure of eating rich foods. Similar conflicts arise between material properties and performance metrics in the choice of materials (Sirisalee et al., 2004).

Among the many fields where MCDM applied (computer software selection, project selection, and system selection), materials selection is certainly one of the most crucial. Searching for suitable materials is a key part of the engineering design process, and is a multi-dimensional problem with many "boxes ticked" at the same time. Changing the materials set in an established technology is a rare event and can be considered as a revolution (Curtarolo et al., 2013). Furthermore, materials selection is the prerequisite for a chain of different engineering selection problems, such as process selection, machine selection, tools selection, material handling equipment selection, supplier selection, and personnel selection (Jahan and Edwards, 2015). Traditionally, choosing a new material or replacing an existing material with another material whose characteristics provide better performance, is usually carried out by applying "trial-and-error" methods and/or by using previous experience. This may or may not result in an optimum design solution but the adoption of MCDM methods will help to avoid the use of inappropriate materials and make sure costs are kept to a minimum.

MCDM address the need for a numerate structure (Charles et al., 1997) in the materials selection process. MCDM provides a foundation for selecting, sorting, and prioritizing materials and help in the overall assessment. The use of MCDM is particularly important when:

- The application is complex or advanced.
- The materials and/or application are new.
- The use of leading edge technology is involved, such as aerospace, electronics, nuclear, and biomedical applications, where product differentiation and competitive advantage are often achievable with just small gains in material performance.

Due to considerable disagreement among members of the engineering design community as to the extent that engineering design involves decision-making and to which classical decision theory applies to engineering, design engineers have not adopted the formalisms of decision theory, despite its long history and wide acceptance in other communities. Therefore, many well accepted methods in engineering design clearly produce questionable results (Hazelrigg, 2003).

Unlike the exact sciences, where there is usually only one single correct solution to a problem, materials selection and substitution decisions require the simultaneous consideration of conflicting advantages and limitations, necessitating compromises and trade-offs; as a consequence, different satisfactory solutions are possible (Farag, 2002). Suppose you need to choose a material for an economic lightweight design from the hypothetical materials shown in Table 4.3. Should the cheapest material or the lightest material be chosen? If Materials A and B are compared, it cannot be determined that either is superior without knowing the relative importance of weight versus price. However, comparing Materials B and C shows that Material C is better than Material B for both objectives, and as a consequence Material C "dominates" Material B. Therefore, as long as Material C is a feasible option, there is no reason that B should be chosen. To conclude the comparisons, it can also be seen that Material D is dominated by Material E. The rest of the material options (Materials A, C and E) have a trade-off associated with weight versus

Table 4.3 **Hypothetical data for materials selection of a lightweight design**

Materials	Weight (kg)	Material price ($)
A	4	850
B	3.6	1000
C	3.2	900
D	3	1150
E	2.4	1100

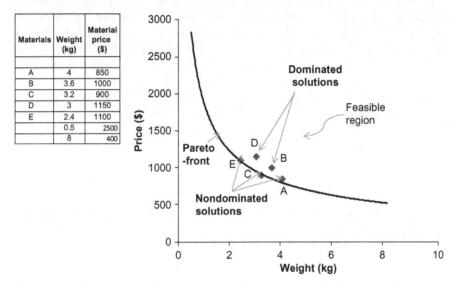

Figure 4.6 The trade-off plot for materials selection of a lightweight design, showing dominated and nondominated solutions.

price, so none is clearly superior to the others. This is called the "nondominated" set of solutions because none of the solutions are dominated. Usually, solutions of this type form a typical shape as shown in Fig. 4.6.

Solutions that lie along the line are nondominated solutions while those that lie inside the line are dominated because there is always another solution on the line that has at least one objective that is better. The line is called the "Pareto-front" and solutions on it are called "Pareto-optimal." All Pareto-optimal solutions are nondominated. Thus, it is important to find the solutions as close as possible to the Pareto-front, as far along it as possible. This is more challenging in MOO problems with constraints that have many solutions in the feasible region. One way of finding points on the Pareto-front is converting all but one into constraints in the modeling phase or invent weightings for the criteria and optimize the weighted sum but this simplifies the consideration and loses information (Miettinen, 1999). There are different methods used in practice for both MADM and MODM problems. In MOO, one method is to use a genetic algorithm (Awad et al., 2012; Cui et al., 2008) to enumerate points along the Pareto-front over several iterations, then use a method to rank the quality of the trade-offs based on the particular application being modeled. Solving MOO problems requires a correct and proper formulation of the problem. In most of the practical optimization problems inaccuracy and uncertainty is present and a clear formulation of the problem may be given after the problem is solved (Miettinen, 1999).

It is observed that there has been a growth in the study of the material evaluation and selection problems using the MCDM approaches from 2006 (Jahan et al., 2010). It is expected that this will keep increasing in the coming years, because in materials selection there are often several solutions for a particular application and materials affect many aspects of a product design, that is, shape, manufacturing process, and product performance, so a more precise approach is required. The "Ashby" materials selection chart method discussed in Chapter 3 does allow a cost-based approach. The method also has the versatility of being able to examine many materials at a glance and allow competing materials to be quickly identified. However, it cannot be guaranteed that the selected material is the best because of the limit of only being able to consider two or three criteria simultaneously, as discussed in the previous section.

4.5 Application of MODM and MADM in material design and selection

The goal of choosing a material is to optimize a number of different metrics of performance for a component in a product in which the material is to be used. Typical metrics include cost, mass, volume, power-to-weight ratio, energy density, etc. but there are many other characteristics that can be in conflict with each other (Ashby, 2000). This must be achieved subject to constraints; that the component can carry the given loads without failure, that certain dimensions are fixed, and that its cost is within certain limits (Weaver et al., 1996).

The concept of material design is not limited to selecting an available material from a database, instead it covers tailoring material structure at various levels of hierarchy (atoms, microstructure, etc.) that are customized for a particular application as shown in Fig. 4.7 (McDowell et al., 2010a). The term materials design may have different meanings to different people (designer, engineer, etc.). A recent definition is (McDowell et al., 2010a): "the top-down driven, simulation-supported, decision-based design of material hierarchy to satisfy a ranged set of product-level performance requirements." The design process is multi-level in the sense that decisions must be made with respect to structure at each level of material and product hierarchies. Multiple levels of models must be integrated with design decisions in concurrent materials/product design (McDowell et al., 2010a).

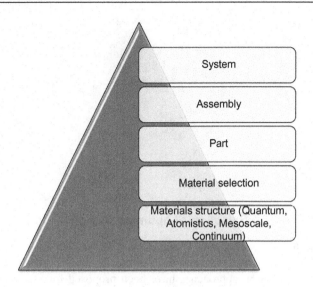

Figure 4.7 Extension of systems-based, top-down materials design from components, subassemblies, assemblies, and components to hierarchical levels of material structure in concurrent materials and product design (McDowell et al., 2010a)

The inherent difficulty with materials selection is choosing a material for an application-specific requirement that may conflict in terms of demands on material structure and properties. For example, a multi-functional material in the mechanical property domain might require target ranges of strength and ductility as conflicting requirements. On the other hand, process lead times and cost for the development of new materials have remained relatively constant due to the "trial-and-error" approach adopted historically by materials engineers and developers (McDowell et al., 2010a).

Optimum design is the selection of the material and geometry to minimize or maximize a given performance metric (Ashby, 2000; Rakshit and Ananthasuresh, 2008). A key objective of mechanical engineering design is to define the dimensions of a component and the materials from which it is made so that it can perform a function acceptably (Edwards, 2005). Edwards (2002, 2003) has pointed out the importance of integrating materials selection with component design. The area of the design decision-making for simultaneous consideration of the structural solution and materials selection, which is generally needed at the early design stage, is relatively weak (Edwards and Deng, 2007). Engineering designers either assume a material before optimizing the geometry or select the best material for an existing geometry of a component or structure but either approach does not guarantee the optimal combination of geometry and material (Rakshit and Ananthasuresh, 2008; Deng and Edwards, 2007). Therefore, engineering designers should consider geometry optimization and materials selection simultaneously (Sepulveda, 1995; Rakshit and Ananthasuresh, 2008; Stolpe and Svanberg, 2004) as shown in Fig. 4.8.

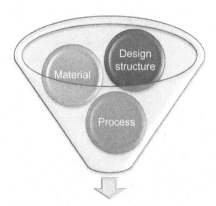

Selection of optimum design

Figure 4.8 Simultaneous consideration of material, structure and process for optimum design.

Extensive optimization approaches have been proposed for material design and selection, and simultaneous optimization of materials and design such as mathematical programming, neural networks and genetic algorithms. The wide variety of material combinations, matrix and reinforcement, geometries, and architectures in the specific case of engineering design of composite materials raised the need for such tools. For example, Cui et al. (2008) in the optimal design of an automotive body assembly formulated the design problem as a multi-objective nonlinear mathematical programming problem with discrete and continuous variables. The discrete variables were the material types and continuous variables were the thicknesses of the panels and the problem was solved using a multi-objective genetic algorithm. An artificial neural network was employed to approximate the constraint functions and reduce the number of finite element runs. Also, concurrent materials selection and structural optimization (geometry and size) based on mixed variables and a genetic algorithm was considered recently by Tang et al.(2010).

An MOO problem with constraints will have many solutions in the feasible region. The optimization of multi-objectives falls into the category of Pareto-optimization, leading to a family of solutions presented to the designer. While at first glance it may appear feasible to reduce MOO problems to single-objective optimization problems via combining nondimensional performance metrics, for example, chart method, one cannot distinguish the sensitivity to variation of individual design variables in such an approach, so treatment of uncertainty is problematic (McDowell et al., 2010c).

In material design and selection problems, the multi-criteria decision-maker can be one person (the designer) or several persons (the design team), the data might be deterministic, stochastic, continuous, or discrete, and objectives and constraints might be linear or nonlinear. MCDM is in the early stages of development in this area of engineering and only practical design studies can speed up the progress. Some developments of MCDM in materials selection and design are described in Chapter 7.

4.6   Utilizing outputs from finite element analysis as inputs to MADM in materials selection

Finite element analysis (FEA) allows material property data to be transmitted directly to a design software package so that the effect of changing material properties on the geometry and dimensions of a component design can be directly evaluated on a computer. At the same time the engineering designer can evaluate the effect of changing geometry and dimensions on product performance. Shanian et al. (2012) recently proposed a combined FEA-MADM approach as a framework that links the capabilities of FEA tools to the MADM approaches for structural materials selection problems. They highlighted that the ability to test preliminary designs is not economically feasible and the assessment of preliminary material systems necessitates the use of numerical prediction tools. Such a combination not only improves multi-criteria materials selection of multi-materials (composites and hybrid materials) (Sirisalee et al., 2006) but also places designers in a position to be able to consider the role of materials selection in the concurrent design of materials and products.

Furthermore, it seems that there are fewer attempts for applying the combined Quality Function Deployment-FEA-MADM-Design of Experiment (DoE) approach for design improvement (Jahan and Bahraminasab, 2015). MADM methods in combination with DoE can enhance the outcomes of computational experiments in which DoE determines important design factors and MADM helps in ranking of design scenarios with diverse qualitative and quantitative data. There appears to be a simulation-based materials design revolution underway in which materials selection is improved by the design of material microstructure and/or mesostructure to satisfy a specified set of performance requirements. Often these multiple performance requirements are in conflict with each other in terms of their demands on microstructure, which leads to the need for multi-objective methods to search for satisfactory solutions (McDowell et al., 2010a). Therefore, significant additional research is needed in these areas to more rapidly qualify new material designs and to shift workload and decision-making based on costly and time-consuming physical experimentation to less costly computational modeling and design (McDowell et al., 2010b). Design decision-making problems need different information, tools, and techniques when it is not possible to find the best design solution based on traditional methods.

4.7   Summary and conclusions

Despite the wide acceptance and long history of the application of Operations Research and decision sciences in other fields, engineering design decisions are mostly limited to unreliable methods. In materials engineering, the use of MCDM is important when the materials are new, the application is complex, new or advanced, or where the product differentiation and competitive advantage are often achievable with just small gains in materials performance. MCDM techniques are divided into MADM and MODM. MADM methods are used for materials selection problems as well as design selection and process selection problems. MODM methods can be fitted to material design and optimization problems as well as simultaneous material and design optimization studies. Although a large amount of research has been done in the recent past on materials selection using different MADM methods, there is still a need to develop MODM methods for use in engineering design applications and based more closely on the true nature of engineering design problems.

Review questions

1. Discuss the implications of having to make compromises on decision-making when selecting suitable materials for a component design.
2. Why is it desirable for the selection of materials and the design of a component to be done simultaneously?
3. Describe the problems associated with selecting available materials to meet specific multiple requirements that conflict with each other.
4. Explain the advantages and disadvantages of using graphical-based methods in materials selection.
5. Explain how materials design can help optimize materials utilization in the design of components.
6. Why is it an advantage to apply MCDM in the engineering design process, particularly when selecting materials?
7. How can MADM and MODM be effectively used in optimal materials selection and material design?
8. What are the benefits of combining decision support and analysis tools in design and materials engineering?

References

Ashby, M.F., 2000. Multi-objective optimization in material design and selection. Acta Mater. 48, 359–369.

Awad, Z.K., Aravinthan, T., Zhuge, Y., Gonzalez, F., 2012. A review of optimization techniques used in the design of fibre composite structures for civil engineering applications. Mater. Des. 33, 534–544.

Charles, J.A., Crane, F.A.A., Furness, J.A.G., 1997. Selection and Use of Engineering Materials. Butterworth-Heinemann, Oxford.

Chiner, M., 1988. Planning of expert systems for materials selection. Mater. Des. 9, 195–203.

Cui, X., Wang, S., Hu, S.J., 2008. A method for optimal design of automotive body assembly using multi-material construction. Mater. Des. 29, 381–387.

Curtarolo, S., Hart, G.L., Nardelli, M.B., Mingo, N., Sanvito, S., Levy, O., 2013. The high-throughput highway to computational materials design. Nat. Mater. 12, 191−201.

Deng, Y.M., Edwards, K.L., 2007. The role of materials identification and selection in engineering design. Mater. Des. 28, 131−139.

Edwards, K.L., 2002. Linking materials and design: an assessment of purpose and progress. Mater. Des. 23, 255−264.

Edwards, K.L., 2003. Designing of engineering components for optimal materials and manufacturing process utilisation. Mater. Des. 24, 355−366.

Edwards, K.L., 2005. Selecting materials for optimum use in engineering components. Mater. Des. 26, 469−473.

Edwards, K.L., Deng, Y.M., 2007. Supporting design decision-making when applying materials in combination. Mater. Des. 28, 1288−1297.

Farag, M.M., 1997. Materials Selection for Engineering Design. Prentice-Hall, New York, NY.

Farag, M.M., 2002. Quantitative methods of materials selection. In: Kutz, M. (Ed.), Handbook of Materials Selection, London, John Wiley & Sons.

Hazelrigg, G.A., 2003. Validation of engineering design alternative selection methods. Eng. Optim. 35, 103−120.

Hwang, C.L., Yoon, K., 1981. Multiple Attribute Decision Making: Methods and Applications: A State-of-the-Art Survey. Springer-Verlag, New York, NY.

Jahan, A., Bahraminasab, M., 2015. Multicriteria decision analysis in improving quality of design in femoral component of knee prostheses: influence of interface geometry and material. Adv. Mater. Sci. Eng. 16.

Jahan, A., Edwards, K.L., 2015. A state-of-the-art survey on the influence of normalization techniques in ranking: Improving the materials selection process in engineering design. Mater. Des. 65, 335−342.

Jahan, A., Ismail, M.Y., Sapuan, S.M., Mustapha, F., 2010. Material screening and choosing methods—a review. Mater. Des. 31, 696−705.

Mcdowell, D.L., Panchal, J.H., Choi, H.-J., Seepersad, C.C., Allen, J.K., Mistree, F., 2010a. Integrated material, product, and process design—a new frontier in engineering systems design. Integrated Design of Multiscale, Multifunctional Materials and Products. Butterworth-Heinemann, Boston, MA (Chapter 1).

Mcdowell, D.L., Panchal, J.H., Choi, H.-J., Seepersad, C.C., Allen, J.K., Mistree, F., 2010b. Critical path issues in materials design. Integrated Design of Multiscale, Multifunctional Materials and Products. Butterworth-Heinemann, Boston, MA (Chapter 2).

Mcdowell, D.L., Panchal, J.H., Choi, H.-J., Seepersad, C.C., Allen, J.K., Mistree, F., 2010c. Overview of the framework for integrated design of materials, products, and design processes. Integrated Design of Multiscale, Multifunctional Materials and Products. Butterworth-Heinemann, Boston, MA (Chapter 3).

Miettinen, K., 1999. Nonlinear Multiobjective Optimization. Kluwer Academic Publishers, Boston, MA.

Rakshit, S., Ananthasuresh, G.K., 2008. Simultaneous material selection and geometry design of statically determinate trusses using continuous optimization. Struct. Multidiscip. Optim. 35, 55−68.

Sepulveda, A.E., 1995. Optimal material selection using branch and bound techniques. AIAA J. 33, 340−347.

Shanian, A., Milani, A.S., 2012. A combined finite element-multiple criteria optimization approach for materials selection of gas turbine components. J. Appl. Mech. 1, 304.

Sirisalee, P., Ashby, M.F., Parks, G.T., Clarkson, P.J., 2004. Multi criteria material selection in engineering design. Adv. Eng. Mater. 6, 84−92.

Sirisalee, P., Ashby, M.F., Parks, G.T., John Clarkson, P., 2006. Multi criteria material selection of monolithic and multi materials in engineering design. Adv. Eng. Mater. 8, 48–56.

Stolpe, M., Svanberg, K., 2004. A stress-constrained truss-topology and material-selection problem that can be solved by linear programming. Struct. Multidiscip. Optim. 27, 126–129.

Tang, X., Bassir, D.H., Zhang, W., 2010. Shape, sizing optimization and material selection based on mixed variables and genetic algorithm. Optim. Eng. 12, 1–18.

Weaver, P.M., Ashby, M.F., Burgess, S., Shibaike, N., 1996. Selection of materials to reduce environmental impact: a case study on refrigerator insulation. Mater. Des. 17, 11–17.

Multi-attribute decision-making for ranking of candidate materials

Learning Aims

The overall aim of this chapter is to describe developed Multi-Attribute Decision-Making (MADM) methods for materials selection. After carefully studying this chapter you should be able to understand:

- The structure of data in matrix based materials selection problems.
- The difference between dimensionless methods used in MADM.
- The steps and techniques for weighting of criteria.
- How to select the ranking method that is best fitted to a particular situation.
- How to find the final ranking of materials when there is inconsistency in the results of different MADM methods.
- How to perform calculations for ranking a database of promising materials using a spreadsheet.

5.1 Rationalization for using multi-attribute decision-making

All products in the marketplace compete to varying degrees on the basis of performance, appearance, price, reliability, safety, maintainability, etc. All of these attributes depend fundamentally on the design of the product (Wallace and Burgess, 1995) and, for essentially mechanical products on the choice of the materials used. The selection of the most appropriate material, or combination of materials, is a demanding intellectual process that takes a lot of time and experience. There are a large number of established and newly developed materials and their associated processes, necessitating the simultaneous consideration of many conflicting criteria.

The large number of current and growing number of new materials coupled with the complex relationships between the different selection parameters, often make the selection of a material for a given component a difficult task (Farag, 2002). Therefore, for contemporary materials selection systems, the suitability of candidate materials is evaluated against multiple criteria and decision-support tools rather than considering a single factor.

Almost always, more than one material is suitable for an engineering application, and the final selection is a compromise that brings some advantages as well as disadvantages. The selection of an optimal material for an engineering design or manufacturing process from among two or more alternative materials on the basis

Multi-criteria Decision Analysis. DOI: http://dx.doi.org/10.1016/B978-0-08-100536-1.00005-9

of two or more attributes is a MADM problem (Rao and Davim, 2008). After identifying the material selection attributes and creating a short-list of materials in a given engineering application, the MADM methods can be used to rank and select the optimum material. The decision variables can be quantitative or qualitative.

5.2   Introduction to the ranking of materials

From the perspective of engineering design and manufacturing, there is a need for optimal selection of materials for reasons of cost reduction, product reliability improvement, manufacturing process yield enhancement, and for replacing obsolete, restricted access, or barred materials (Michael, 2009b). The selection of the most appropriate material, or combination of materials, is a demanding intellectual process that takes a lot of time and experience, and it is associated with all but the simplest of design and manufacturing problems (Ullah and Harib, 2008). An inappropriate selection of materials may result in damage or failure of a component or an assembly of components and significantly decrease the performance. Considerable technical and commercial benefits can therefore be gained by the use of appropriate material(s), compensating for the time taken in their identification. Fig. 5.1 shows that ranking and selection of the optimal material is a key stage in the materials selection process.

Table 5.1 shows the result of using the Simple Additive Weighting (SAW) method for ranking of hypothetical candidate materials for the wing spar of an aircraft. It appears that the method is not reliable because Material 1 is not appropriate in practice for lightweight structures. Simanaviciene and Ustinovichius (2010) showed that the Technique for Order of Preference by Similarity to Ideal Solution (TOPSIS) is more sensitive than the SAW method to the values of criteria.

The Limits on Properties (LOP) approach (Farag, 1979, Farag, 1997; Dieter, 1997), which is similar to SAW, has traditionally been used for ranking of

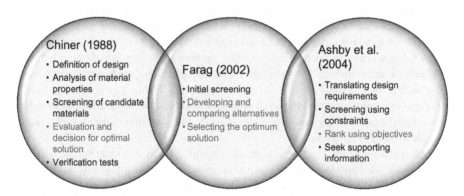

Figure 5.1 Ranking in materials selection steps (Chiner, 1988; Farag, 2002; Ashby et al., 2004).

Table 5.1 **Normalized values (higher is better) of hypothetical data for the wing spar structure of an aircraft and ranking results using SAW method**

	Density	Creep resistance	Compressive strength	Price	Results
Weighting	0.4	0.2	0.2	0.2	
Material 1	1	10	10	10	6.4
Material 2	5	5	5	5	5
Material 3	10	1	1	1	4.6

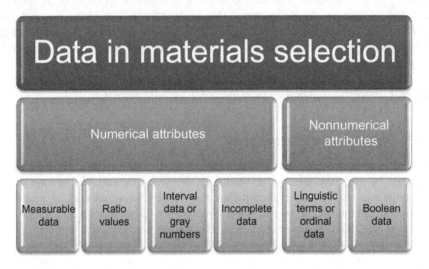

Figure 5.2 Structure of data in materials selection.

materials, although recently it has been shown that the method suffers from short-comings (Dehghan-Manshadi et al., 2007; Jahan et al., 2011b; Jahan et al., 2012a).

5.3   Structure of data in materials selection

Materials can have many dimensions: technical, economic, aesthetic, personality, and ecological dimensions (Ashby and Johnson, 2010c). A set of N properties of materials is typically extracted whether obtained by experiment or simulation, and is then related to M performance requirements, comprising a property-performance space of dimension M + N. Data for each dimension have their own characteristics and can be represented in different ways. As a consequence, various information may be provided in materials selection decision matrices as shown in Fig. 5.2 (Ashby et al., 2004).

5.3.1 Numerical attributes

Numerical values express the absolute measure of a material property such as density, modulus, strength and thermal conductivity. The values that state a relative measure of one material with respect to another material are called ratio values (e.g., cost). The values of material properties are often imprecisely measured using ranges (Liao, 1996). Due to the stochastic nature of material processing operations, such properties are usually not fixed but range between two values. For example, the tensile strength of Ti-6V-4Al alloy, cobalt-chromium alloy, and austenitic stainless steel are approximately [750, 1050], [450, 1000], and [200, 1100] MPa, respectively. It is important to determine the maximum and minimum values for a specific property to decide whether the material will meet a specific design requirement for all samples of the material (Michael, 2009a). Therefore, material and mechanical engineers must carefully specify the range of materials' properties and avoid reporting incomplete data. There is the risk that the final choice of material is not accurate because of incomplete, approximate, and possibly incorrect information (Edwards, 2005). The research to date has tended to focus on fixed values of materials' properties rather than on ranges of materials' properties.

5.3.2 Nonnumerical attributes

Some properties are not so easily expressed numerically but they can instead be described by linguistic terms or ordinal data. A "linguistic variable" always exists in criteria such as resistance to corrosion, that can be expressed only as a ranking (e.g., poor, adequate, good), and some can only be captured in text and images (Ashby et al., 2004). Some properties are Boolean in nature such as the ability of a material to be recycled (Ashby et al., 2004). It therefore appears that the final choice of material is at risk because of incomplete, approximate, and possibly incorrect information (Edwards, 2005).

When both tangible and intangible factors need to be considered, a number of people are required to participate in a judgmental process. In order to overcome this problem, group MADM methods have been applied to determine trade-off among criteria for optimal decision-making in materials selection. Recently, Jeya Girubha and Vinodh (2012) have applied fuzzy VIKOR in materials selection to deal with linguistic terms from a group of people or Decision-Makers (DMs), for example, designers, materials engineers, etc. For situations where only one person or DM, for example, a designer or materials engineer, determines the linguistic labels for criteria, the corresponding fuzzy numbers can be obtained using an 11-point conversion scale (Rao and Patel, 2010).

Industrial designers also deal with the personality of materials and this type of information is usually obtained from conducting interviews (Ashby and Johnson, 2010a). A material might express characteristics such as shiny, smooth, transparency, or sexy but these expressive characteristics or meanings are not part of a materials' physical entity or appearance. Furthermore, for example when designers want to convey a meaning of toy-like in a product, plastics and rubber can be candidate materials which associate with sensorial properties like cold, smooth, light, elastic, and ductile, while usually sensorial property represents via some point scales. Generally, these sensorial, invisible, and qualitative properties are called intangible characteristics (Karana et al., 2008).

5.4 Normalization of criteria in MADM

MADM consists of generating alternatives, establishing criteria (attributes), evaluating alternatives, assessing criteria weightings, and the application of a ranking system (Vincke, 1992). Each of the criteria is related to an objective in the given decision context, and normalization is used for transforming different criteria into a compatible measurement. Therefore, normalization of decision matrix elements is a crucial step in most MADM techniques. In addition, it is the first step in some objective weighting methods (Section 5.5), which try to define the importance of criteria only based on the known data of the problem, and are useful when a DM is not involved. If cost, benefit, and target criteria are necessary for a MADM problem, attention should be paid to the normalization techniques that address all types of criteria. In MADM techniques that require significant user interaction during the problem solving process, normalization is usually not necessary. However, when dealing with a database of alternatives, user interaction methods are useless and it is important to empower decision-support systems for materials selection with accurate and inclusive models. It is shown that although many normalization methods may appear to be minor variants of each other, these nuances can have important consequences in engineering design decision-making.

5.4.1 Principal attributes of normalization techniques

This section describes the aspects considered important for evaluation/developing a normalization method as shown in Fig. 5.3 (Jahan and Edwards, 2015). Suppose a decision

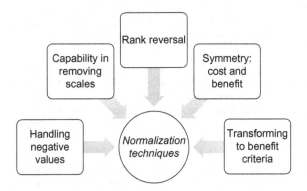

Figure 5.3 Expected properties for normalization methods.

matrix for materials selection includes the following parts: alternatives A_i $(i = 1,\ldots,m)$, which DMs have to choose, criteria C_j $(j = 1,\ldots,n)$, and a decision matrix with x_{ij} elements as shown in Table 4.1. The performance ratings (x_{ij}) for different criteria are measured with different units but in order for the decision matrix to have a valid comparison of all of the elements must be dimensionless. When using selection methods with a matrix-based structure, a proper normalization method can enhance the effectiveness of the final decision. This is suitable for quantitative criteria but for qualitative criteria it is important to bear in mind that any scoring system used, for example, five-point versus three-point (Jacobs et al., 2014), might also affect the final ranking.

- *Capability in removing scales*

    It is a basic rule that when normalizing identical data with different units or scales, the same results are obtained. The same criterion function can be demonstrated using different "convertible" units, for example, density can have units [kg/m^3] or [g/cm^3]. For temperature, the Fahrenheit (T_F) scale is used in the United States but most of the rest of the world uses the Celsius (T_C) scale, and in science it is often more convenient to use the Kelvin (T_K) scale. These "convertible" units affect material properties such as heat transfer coefficient or thermal conductivity, with temperature being related as follows:

$$T_F = \frac{9}{5}T_C + 32 \quad T_C = T_K - 273$$

Normalization methods must be able to eliminate the "convertible" units of criterion functions, $(\varphi_{ij} = \alpha r_{ij} + \beta)$, and return the same result for all scales (Opricovic and Tzeng, 2004).

- *Symmetry in normalization for cost and benefit criteria*

    In order to preserve the maximum initial information in relation to initial attribute values and values of other criteria, it is necessary to check the symmetry of normalized values when comparing cost and benefit criteria. For example, in some available normalization methods, the benefit criteria can be normalized in the interval [p, 1], $(0 < p < 1)$, while in case of cost type criteria that value belongs either in the interval from "0" to "p" or "0" to "1." It can be a type of asymmetry that some methods cannot cover the whole range of "0" to "1." Another type of asymmetry appears in normalized values of the same data when it is normalized as cost and benefit criteria.

- *Transforming other types of criteria to the benefit criteria*

    Some methods just remove the scale of criteria but cannot convert cost criteria to benefit criteria. Although such transformation is not necessary for all MADM methods, it usually reduces the calculations required in the subsequent stages.

- *Rank reversals*

    In this phenomenon, the alternatives' order of preference (ranking) changes when an alternative is added to or removed from the decision problem. In addition, the number of alternatives will affect the normalization process. In some cases this may lead to what is called total rank reversal, where the order of preferences is totally inverted. The reason for rank reversal can be the use of an improper normalization technique.

- *Handling negative values*

    Some criteria in a decision matrix include negative values, for example, the problem of differences in thermal expansion coefficient of materials chosen to repair aerospace structures (Fouladi et al., 2010), and it is important that the MADM method is able to analyze data in such situations and present acceptable ranking.

5.4.2 Classifying the dimensionless methods

The normalization techniques are classified in this section, with an explanation of the main disadvantages of each group of methods. Table 5.2 shows the sum-based dimensionless methods available for cost and benefit criteria. The properties whose higher values are desirable, called positive criteria or beneficial attributes, and those properties whose smaller values are always preferable, named negative criteria, cost criteria, or nonbeneficial attributes, are presented.

The main weak point of normalization methods listed in Table 5.2 is inefficiency on removing the scales, which means normalized values of x_{ij} $i = 1, 2, 3, \ldots .m$ are different from normalized values of criterion functions ($\varphi_{ij} = \alpha r_{ij} + \beta$). Also some of the methods (e.g., Norm (1) and Norm (2)) suffer from lack of symmetry in the normalization process for cost and benefit criteria. In all the norms listed in Table 5.2, the number of alternatives influences the result of normalization, and it is expected to see rank reversal by adding/removing alternatives. Also, when a decision matrix includes negative performance values (x_{ij}), some methods (e.g., Norms (3) and (4)) are not useful.

Table 5.3 shows the linear ratio-based normalization methods; same as sum-based normalization methods, the main shortcoming of these techniques is the inability for removing scales. As explained earlier, selecting the wrong pairs for cost and benefit criteria cause asymmetry in normalization process, while it has not taken to account in many studies. It was shown that the best pair for Norm (9) is Norm (12).

Table 5.4 demonstrates linear max−min normalization methods. Among these methods, Norms (16) and (17) are able to remove convertible units ($\alpha r_{ij} + \beta$). However, other methods listed in Table 5.4 do not have such an advantage. It seems the only disadvantage of Norms (16) and (17) are that the number of alternatives influences the result of normalization, because adding or removing alternatives probably changes the maximum and minimum values.

Table 5.2 Sum-based dimensionless methods in MADM for cost and benefit criteria

Number	Condition of use	Formula	Number	Condition of use	Formula
Norm (1)	Benefit criteria	$r_{ij} = \dfrac{x_{ij}}{\sqrt{\sum\limits_{i=1}^{m} x_{ij}^2}}$	Norm (5)	Benefit criteria	$r_{ij} = \dfrac{\ln(x_{ij})}{\ln\left(\prod_{i=1}^{m} x_{ij}\right)}$
Norm (2)	Cost criteria	$r_{ij} = 1 - \dfrac{x_{ij}}{\sqrt{\sum\limits_{i=1}^{m} x_{ij}^2}}$	Norm (6)	Cost criteria	$r_{ij} = \dfrac{1 - \frac{\ln(x_{ij})}{\ln(\prod_{i=1}^{m} x_{ij})}}{m - 1}$
Norm (3)	Benefit criteria	$r_{ij} = \dfrac{x_{ij}}{\sum\limits_{i=1}^{m} x_{ij}}$	Norm (7)	Benefit criteria	$r_{ij} = 1 - \dfrac{x_j^{\max} - x_{ij}}{\sum\limits_{i=1}^{m}(x_j^{\max} - x_{ij})}$
Norm (4)	Cost criteria	$r_{ij} = \dfrac{1/x_{ij}}{\sum\limits_{i=1}^{m} 1/x_{ij}}$	Norm (8)	Cost criteria	$r_{ij} = 1 - \dfrac{x_{ij} - x_j^{\min}}{\sum\limits_{i=1}^{m}(x_{ij} - x_j^{\min})}$

Table 5.3 **Linear ratio-based normalization methods for cost-benefit criteria**

Number	Condition of use	Formula
Norm (9)	Benefit criteria	$r_{ij} = \dfrac{x_{ij}}{x_j^{\max}}$
Norm (10)	Cost criteria	$r_{ij} = 1 - \dfrac{x_{ij}}{x_j^{\max}}$
Norm (11)		$r_{ij} = \dfrac{x_j^{\min}}{x_{ij}}$
Norm (12)		$r_{ij} = 1 - \dfrac{x_{ij} - x_j^{\min}}{x_j^{\max}}$
Norm (13)		$r_{ij} = \dfrac{1/x_{ij}}{1/x_j^{\max}} = \dfrac{x_j^{\max}}{x_{ij}}$
Norm (14)	Benefit criteria	$r_{ij} = \left(\dfrac{x_{ij}}{x_j^{\max}}\right)^2$
Norm (15)	Cost criteria	$r_{ij} = \left(\dfrac{x_j^{\min}}{x_{ij}}\right)^3$

Table 5.4 **Linear max−min dimensionless methods for cost and benefit criteria**

Number	Condition of use	Formula
Norm (16)	Benefit criteria	$r_{ij} = \dfrac{x_{ij} - x_j^{\min}}{x_j^{\max} - x_j^{\min}}$
Norm (17)	Cost criteria	$r_{ij} = \dfrac{x_j^{\max} - x_{ij}}{x_j^{\max} - x_j^{\min}}$
Norm (18)	Benefit criteria	$r_{ij} = \dfrac{x_{ij}}{x_j^{\max} - x_j^{\min}}$
Norm (19)	Cost criteria	$r_{ij} = \dfrac{x_{ij}}{x_j^{\min} - x_j^{\max}}$
Norm (20)	Benefit criteria	$r_{ij} = 1 - \left\lvert\dfrac{x_j^{\max} - x_{ij}}{x_j^{\max}}\right\rvert$
Norm (21)	Cost criteria	$r_{ij} = 1 - \left\lvert\dfrac{x_j^{\min} - x_{ij}}{x_j^{\min}}\right\rvert$

The dimensionless methods proposed for beneficial and nonbeneficial criteria in materials selection are shown in Table 5.5. Dehghan-Manshadi et al. (2007) developed a nonlinear normalization model (Norms (22) and (23)) for benefit and cost criteria, for which the critical values need to be specified by the engineering designer. The method normalizes data between −100 and 100. Fayazbakhsh et al. (2009) proposed Z transformation (Norms (24) and (25)) for dimensionless of decision matrix in weighted properties method (WPM) or SAW but the output of this normalization technique is around zero and contains negative values. Therefore, there are some limitations in both methods when applied to other MADM methods.

Table 5.5 Normalization method proposed in materials selection for cost and benefit criteria

Number	Condition of use	Formula	
Norm (22)	Benefit criteria	$Y = a_1\ln(b_1X + c_1)$ for $X_c \neq \dfrac{X_{max}}{2}$ $Y = \dfrac{200X}{X_{max}} - 100$ for $X_c = \dfrac{X_{max}}{2}$	$a_1 = \dfrac{-100}{\ln(X_c/(X_{max} - X_c))}$, $b_1 = \dfrac{X_{max} - 2X_c}{X_c(X_{max} - X_c)}$, $c_1 = \dfrac{X_c}{X_{max} - X_c}$
Norm (23)	Cost criteria	$Y = a_2\ln\left(\dfrac{b_2}{X} + c_2\right)$ for $X_c \neq 2X_{min}$ $Y = \dfrac{200X_{min}}{X} - 100$ for $X_c = 2X_{min}$	$a_2 = \dfrac{-100}{\ln(-X_{min}/(X_{min} - X_c))}$, $b_2 = \dfrac{-X_c^2 + 2X_{min}X_c}{X_{min} - X_c}$, $c_2 = \dfrac{-X_{min}}{X_{min} - X_c}$
Norm (24)	Benefit criteria	$\mu_j = \dfrac{\sum_{j=1}^{m} x_{ij}}{m}$, $\sigma_j = \sqrt{\dfrac{\sum_{j=1}^{m}(x_{ij} - \mu_j)^2}{m}}$	$r_{ij} = \dfrac{x_{ij} - \mu_j}{\sigma_j}$
Norm (25)	Cost criteria		$r_{ij} = -\dfrac{x_{ij} - \mu_j}{\sigma_j}$

Table 5.6 Normalization methods for all kinds of criteria

Number	Formula	
Norm (26)	$r_{ij} = \dfrac{|x_{ij} - T_j|}{x_{ij}^{max} - T_j}$	
Norm (27)	$r_{ij} = \begin{cases} \dfrac{x_j^{min}}{x_{ij}} & \text{For cost criteria} \\[2ex] \dfrac{x_{ij}}{x_j^{max}} & \text{For benefit criteria} \\[2ex] \dfrac{\min\{x_{ij}, T_j\}}{\max\{x_{ij}, T_j\}} & \text{For target criteria} \end{cases}$	
Norm (28)	$r_{ij} = e^{\frac{(x_{ij}-T_j)^2}{-2\sigma_j^2}}$	
Norm (29)	$r_{ij} = e^{\left(\frac{|x_{ij}-T_j|}{\min\{x_j^{min},T_j\}-\max\{x_j^{max},T_j\}}\right)}$	
Norm (30)	$r_{ij} = 1 - \dfrac{|x_{ij} - T_j|}{\max\{x_{ij}^{max}, T_j\} - \min\{x_{ij}^{min}, T_j\}}$	

Table 5.6 shows the normalization methods that can be used for all kinds of criteria including benefit, cost, and target values. Target setting and use is a natural part of decision-making in engineering design problems. There are different published examples of target criteria used in materials and engineering design selection problems (Bahraminasab and Jahan, 2011; Cavallini et al., 2013; Jahan et al., 2011b; Jahan et al., 2012a; Bahraminasab et al., 2014).

Norm (26), or nominal-is-best method, was proposed by Wu (2002). One of the shortcomings of this method is that it will not work out when the target value is higher than the maximum performance rating. Another method called linear method-ideal (Zhou et al., 2006) (Norm (27)) cannot produce symmetric normalized values around the most favorable value. Also, it has all the weak points of the techniques listed in Table 5.3. Jahan et al. (2011b) proposed Norm (29), which uses an exponential function to cover all types of criteria and address the shortcomings of LOP method (Farag, 1979; Dieter, 1997). Jahan et al. (2012a) also proposed Norm (30), which is called the target-based normalization method. The normalized data must be commensurate with the raw data. Nonlinear functions provide an alternative here. It must, however, be noticed that every nonlinear function distorts the original problem. An advantage of target criteria is that they convert all types of data to the benefit criteria.

The accuracy of the normalization process is very significant in the case of engineering design decision-making problems because it must address diverse criteria and objectives.

5.5 Weighting procedure of criteria

A lot of multi-criteria decision-making methods (MCDM) have been proposed to address materials selection problems (Jahan et al., 2010). Many of these methods require quantitative weightings for the attributes. Since weighting of attributes play a very significant role in the ranking of results of the alternatives, therefore one crucial problem is to assess the weightings or relative importance of materials properties (Diakoulaki et al., 1995). Furthermore, the reasonableness of the weighting assignment has an important impact on the reliability and accuracy of the decision results (Rongxi et al., 2009).

Weighting methods that try to define the importance of the criteria can be categorized into three groups (Fig. 5.4): (1) subjective methods in which the role of assigning the importance to the criteria is put on the shoulders of the DM or designer, (2) objective methods in which DM has no role in determining the importance of the criteria, and (3) the combined weighting scheme of the two previous groups.

5.5.1 Subjective methods

The subjective methods of determining the weighting of attributes is based solely on preference information of attributes given by expert evaluation and can be according to previous experience, particularly the constraints of design (Jee and Kang, 2000), or designers' preferences (Shanian and Savadogo, 2006d). These methods can be categorized as follows.

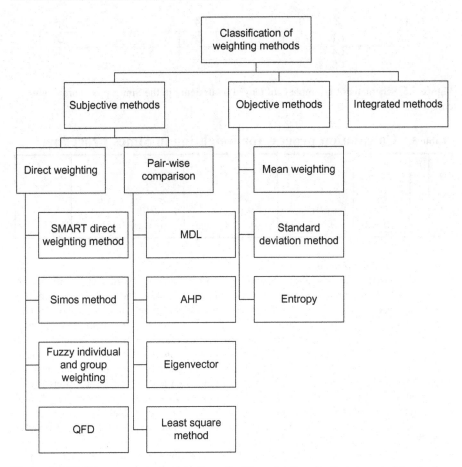

Figure 5.4 Well-known methods typically applied for weighting of criteria in materials selection.

5.5.1.1 Direct weighting procedure

• *SMART direct weighting method*

There are numerous techniques to directly determine the subjective weightings. These include SWING (Von Winterfeldt and Edwards, 1986), TRADEOFF (Keeney and Raiffa, 1993; Pöyhönen and Hämäläinen, 2001), direct rating (Von Winterfeldt and Edwards, 1986; Bottomley and Doyle, 2001), point allocation (PA) (Doyle et al., 1997), Delphi method (Hwang and Lin, 1987), and Simple Multi-attribute Rating Technique (SMART) (Edwards, 1977; Ward and Hutton, 1994). In these methods, the DM allocates numbers to directly describe the weightings of the attributes. For instance in SMART, attributes are first ranked based on importance, and then rated relative to the least important. Usually, giving ratings begins with assigning 10 points to the least important attribute. The relative importance of the other attributes is then evaluated by giving them points from 10 upwards. The research of Pöyhönen and Hämäläinen (2001) showed that in DIRECT, SWING, and TRADEOFF methods, weightings do not differ from each other.

Figure 5.5 Schematic of a sample card played by designer in the Simos procedure (eight criteria).

Table 5.7 Calculation process for weighting in Simos procedure

r	1	2	3	4	5	$S = \sum\limits_{r=1}^{4} S_r$
S'_r	1	0	2	3		
S_r	2	1	3	4		10
P_r	1	1 + 9/10 (0 + 2)	1 + 9/10 (0 + 2 + 1)	1 + 9/10 (0 + 2 + 1 + 3)	1 + 9/10 (0 + 2 + 1 + 3 + 4)	

- *The revised Simos method*

 The revised Simos' weighting method, which is based on a "card playing" procedure, was used first in materials selection by Shanian et al. (2008). The method is simple and practical but it occasionally leads to the same weightings in an uncontrolled manner (Figueira and Roy, 2002). Each (material property) criterion's name should be written on a card and you are asked to arrange the cards from the least to the most important (such as a horizontal scale from left to right as shown schematically in Fig. 5.5). Of course some criteria may be of the same importance (i.e., same weighting) to you, and you can put them in the same level/subset. The distance between any two adjacent subsets is considered to be identical and equal to a scale unit. Each two successive subsets can then be distinguished further by adding blank cards among them. Each blank card placed by the designer between two successive subsets means one more scale unit difference between their corresponding weightings.

 To illustrate Fig. 5.5, C3 and C2 are the least and the most important criteria, respectively. C1, C4, and C6 are in the same level, which means they have same importance. Further, the two blank cards say between the C8 and C5 indicates two more scale unit difference between the C8 and C5 and C7 weightings.

 It is additionally required that you give an estimation of the ratio of the most to the least important criterion. This ratio is denoted by z ($z \geq 1$). Eq. (5.1) defines the number of scale units. The numbers of subsets in Fig. 5.5 equals five ($\bar{n}$). Table 5.7 shows the calculation process for the example described in Fig. 5.5.

$$S_r = S'_r + 1; \quad r = 1, 2, 3, \ldots, \bar{n} - 1 \tag{5.1}$$

where S'_r is the number of blank cards between the rth and $r + 1$th subsets.

 Note that if no blank card is used between two successive subsets, there exists one unit difference between their positions on the scale. The total number of units between the first and the last subsets, S, can be calculated by Eq. (5.2).

$$S = \sum_{r=1}^{4} S_r \tag{5.2}$$

Table 5.8 **Normalized weightings in the Simos procedure**

Criteria	C3	C1	C4	C6	C8	C5	C7	C2	$P' = \sum_{j=1}^{8} P'_j$
j	1	2	3	4	5	6	7	8	
P_r	P_1	P_2	P_2	P_2	P_3	P_4	P_4	P_5	
P'_j	1	2.8	2.8	2.8	3.7	6.4	6.4	10	35.9
P^*_j	2.8	7.8	7.8	7.8	10.30	17.82	17.82	27.85	100

As can be seen in Table 5.7, S is 8. Next, based on the definition of the z-ratio in the previous step, the length of the scale unit is obtained by Eq. (5.3). In this example, it is supposed that z equals 10, and u would therefore be 9/10.

$$u = \frac{z-1}{s} \tag{5.3}$$

Eq. (5.4) is used for calculating the nonnormalized weighting of the rth subset.

$$P(r) = 1 + u(S_0 + \cdots + S_{r-1}) \quad r = 1, 2, \ldots, \bar{n}, \quad S_0 = 0 \quad P(r) = 1 \tag{5.4}$$

For each criterion, the nonnormalized weighting P'_j is defined to be the same as the weighting of the subset it belongs to (i.e., within each subset $P'_j = P(r)$). As shown in Table 5.8, the normalized criteria weightings $P^*_j (j = 1, 2, \ldots, n)$ are obtained by Eq. (5.5).

$$P^*_j = \frac{P'_j}{P'} \times 100 \tag{5.5}$$

where

$$P' = \sum_{j=1}^{8} P'_j$$

Generally, the P^*_j values are not whole numbers and rounding off some of their decimal figures may cause the summation of normalized weightings to be smaller than 100. If enough decimal figures are chosen, and the number of criteria is relatively small, the error introduced by rounding off decimal figures is negligible ($\sum_{j=1}^{n} P^*_j \cong 100$).

• *Fuzzy individual and group weighting*

Linguistic variables can be used for calculating the importance of the criteria. It is possible to calculate the weighting of criteria according to the idea of a designer (individual decision-making) or design team (group decision-making). Trapezoidal fuzzy numbers form the most common division of fuzzy numbers and the commonly used triangular fuzzy number is a special case of trapezoidal fuzzy number. Trapezoidal fuzzy numbers can encompass more uncertainty than the triangular fuzzy number. It is used in modeling linear uncertainty in scientific and applied engineering problems (Jeya Girubha and Vinodh, 2012).

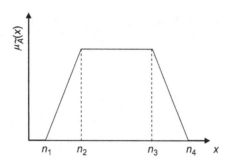

Figure 5.6 Trapezoidal fuzzy number.

Table 5.9 Linguistic terms and corresponding fuzzy numbers for each criterion

Linguistic variable	Fuzzy number
Very poor (VP)	(0.0, 0.0, 0.1, 0.2)
Poor (P)	(0.1, 0.2, 0.2, 0.3)
Medium poor (MP)	(0.2, 0.3, 0.4, 0.5)
Fair (F)	(0.4, 0.5, 0.5, 0.6)
Medium good (MG)	(0.5, 0.6, 0.7, 0.8)
Good (G)	(0.7, 0.8, 0.8, 0.9)
Very good (VG)	(0.8, 0.9, 1.0, 1.0)

A trapezoidal fuzzy number can be defined as $\{(n_1, n_2, n_3, n_4)|n_1, n_2 n_3, n_4 \in R; \quad n_1 \leq n_2 \leq n_3 \leq n_4\}$, which respectively denotes the smallest possible, most promising, and largest possible values, and the membership function is defined using Eq. (5.6) and is shown in Fig. 5.6.

$$\mu_{\tilde{A}}(x) = \begin{cases} \dfrac{x - n_1}{n_2 - n_1} & , x \in [n_1, n_2] \\ 1, & , x \in [n_2, n_3] \\ \dfrac{n_4 - x}{n_4 - n_3} & , x \in [n_3, n_4] \\ 0 & otherwise \end{cases} \qquad (5.6)$$

The linguistic terms and corresponding fuzzy number that uses for defining the importance of criteria by the DM are shown in Tables 5.9.

The linguistic variables and corresponding fuzzy set values for selection of materials for an automotive vehicle instrument panel (Jeya Girubha and Vinodh, 2012) is shown in Tables 5.10 and 5.11. Let the fuzzy rating for the importance weighting of the kth DM be $W_{jk}\{W_{jk1}; W_{jk2}; W_{jk3}; W_{jk4}\}$.

Table 5.10 Importance weighting of criteria assessed by DMs (linguistic variable)

	D1	D2	D3	D4	D5
C1 (Max. temp. limit)	G	G	VG	G	MG
C2 (Recyclability)	G	G	MG	VG	G
C3 (Elongation)	G	G	MG	MG	G
C4 (Weight)	G	MG	VG	G	G
C5 (Thermal conductivity)	G	VG	G	VG	G
C6 (Tensile strength)	G	MG	G	MG	MG
C7 (Cost)	G	G	G	VG	G
C8 (Toxicity level)	MG	G	MG	MG	G

Table 5.11 Importance weighting of materials selection criteria assessed by designers (fuzzy set)

	D1	D2	D3	D4	D5
C1 (Max. temp. limit)	(0.7, 0.8, 0.8, 0.9)	(0.7, 0.8, 0.8, 0.9)	(0.8, 0.9, 1, 1)	(0.7, 0.8, 0.8, 0.9)	(0.5, 0.6, 0.7, 0.8)
C2 (Recyclability)	(0.7, 0.8, 0.8, 0.9)	(0.7, 0.8, 0.8, 0.9)	(0.5, 0.6, 0.7, 0.8)	(0.8, 0.9, 1, 1)	(0.7, 0.8, 0.8, 0.9)
C3 (Elongation)	(0.7, 0.8, 0.8, 0.9)	(0.7, 0.8, 0.8, 0.9)	(0.5, 0.6, 0.7, 0.8)	(0.5, 0.6, 0.7, 0.8)	(0.7, 0.8, 0.8, 0.9)
C4 (Weight)	(0.7, 0.8, 0.8, 0.9)	(0.5, 0.6, 0.7, 0.8)	(0.8, 0.9, 1, 1)	(0.7, 0.8, 0.8, 0.9)	(0.7, 0.8, 0.8, 0.9)
C5 (Thermal conductivity)	(0.7, 0.8, 0.8, 0.9)	(0.8, 0.9, 1, 1)	(0.7, 0.8, 0.8, 0.9)	(0.8, 0.9, 1, 1)	(0.7, 0.8, 0.8, 0.9)
C6 (Tensile strength)	(0.7, 0.8, 0.8, 0.9)	(0.5, 0.6, 0.7, 0.8)	(0.7, 0.8, 0.8, 0.9)	(0.5, 0.6, 0.7, 0.8)	(0.5, 0.6, 0.7, 0.8)
C7 (Cost)	(0.7, 0.8, 0.8, 0.9)	(0.7, 0.8, 0.8, 0.9)	(0.7, 0.8, 0.8, 0.9)	(0.8, 0.9, 1, 1)	(0.7, 0.8, 0.8, 0.9)
C8 (Toxicity level)	(0.5, 0.6, 0.7, 0.8)	(0.7, 0.8, 0.8, 0.9)	(0.5, 0.6, 0.7, 0.8)	(0.5, 0.6, 0.7, 0.8)	(0.7, 0.8, 0.8, 0.9)

The aggregated fuzzy weighting W_j of each criterion is calculated using Eq. (5.7):

$$W_j = \{W_{J1}, W_{J2}, W_{J3}, W_{J4}\} \tag{5.7}$$

where

$$W_{J1} = \min_k\{W_{jk1}\}, \quad W_{J2} = \frac{1}{k}\sum W_{jk2}, \quad W_{J3} = \frac{1}{k}\sum W_{jk3}, \quad W_{J4} = \max_k\{W_{jk4}\}$$

Table 5.12 Aggregated fuzzy values, defuzzied weighting, and final weighting of criteria

	$\{W_{J1}, W_{J2}, W_{J3}, W_{J4}\}$	Defuzzied weighting	Final weighting (normalized)
C1 (Max. temp. limit)	(0.5, 0.78, 0.82, 0.1)	0.77	0.06
C2 (Recyclability)	(0.5, 0.78, 0.82, 1)	0.77	0.06
C3 (Elongation)	(1.25, 1.80, 1.90, 2.25)	1.79	0.15
C4 (Weight)	(0.5, 0.78, 0.82, 1)	0.77	0.06
C5 (Thermal conductivity)	(1.75, 2.10, 2.20, 2.50)	2.13	0.18
C6 (Tensile strength)	(1.25, 1.70, 1.85, 2.25)	1.76	0.15
C7 (Cost)	(1.75, 2.05, 2.10, 2.50)	2.11	0.18
C8 (Toxicity level)	(1.25, 1.70, 1.85, 2.25)	1.76	0.15

The criterion fuzzy weightings are defuzzied using the Centroid method as explained in Eq. (5.8) (Shemshadi et al., 2011) to get Crisp values. The attained Crisp values are shown in Table 5.12.

$$
\begin{aligned}
Defuzz(W_j) &= \frac{\int \mu(w) \cdot w dw}{\int \mu(w) dw} \\[2ex]
&= \frac{\int_{w_{j1}}^{w_{j2}} \left(\dfrac{w - w_{j1}}{w_{j2} - w_{j1}}\right) \cdot w dw + \int_{w_{j2}}^{w_{j3}} w dw + \int_{w_{j3}}^{w_{j4}} \left(\dfrac{w_{j4} - w}{w_{j4} - w_{j3}}\right) \cdot w dw}{\int_{w_{j1}}^{w_{j2}} \left(\dfrac{w - w_{j1}}{w_{j2} - w_{j1}}\right) dw + \int_{w_{j2}}^{w_{j3}} dw + \int_{w_{j3}}^{w_{j4}} \left(\dfrac{w_{j4} - w}{w_{j4} - w_{j3}}\right) dw} \\[2ex]
&= \frac{- w_{j1} w_{j2} + w_{j3} w_{j4} + \dfrac{1}{3}(w_{j4} - w_{j3})^2 - \dfrac{1}{3}(w_{j2} - w_{j1})^2}{- w_{j1} - w_{j2} + w_{j3} + w_{j4}} \\[2ex]
&= \frac{w_{j3}^2 + w_{j4}^2 + w_{j3} \cdot w_{j4} - w_{j1}^2 - w_{j2}^2 - w_{j1} \cdot w_{j2}}{3(- w_{j1} - w_{j2} + w_{j3} + w_{j4})}
\end{aligned}
$$

$$(5.8)$$

5.5.1.2 Pair-wise comparison

In the pair-wise comparison methods, participants are presented with a worksheet and asked to compare the importance of two criteria at a time. Then, the relative importance is scored and the results are normalized to a total of 1.0. This method has the benefit of being easy to calculate. The results are clear and especially distinctive for issues about qualitative factors, which are used for decision-making or evaluation. The pair-wise comparison methods include Analytic Hierarchy Process (AHP) (Saaty, 1990; Rao and Davim, 2008; Rao, 2008), Digital Logic (DL) approach (Farag, 1997), Modified Digital Logic (MDL) approach (Dehghan-Manshadi et al., 2007), eigenvector (Saaty, 1977), and weighted least-square method (Chu et al., 1979). The last two methods allow for the calculation of the attributes' weightings, while there is inconsistency in the DM's idea of pair-wise comparison. Among these methods, DL and MDL enjoy a wide acceptance in materials selection. According to Pöyhönen and Hämäläinen (2001), the inconsistency in AHP depends on the applied evaluation scale and it increases either by higher number of attributes or judging the importance degree. Furthermore, Shirland et al. (2003) used goal programming as a mathematical programming model to determine the weightings based on triad comparison of the attributes.

5.5.2 Objective methods

The objective methods obtain the weightings based only on the known data of the problem. Therefore, objective weighting methods would be useful when a DM is nonexistent. Also, the objective weighting is particularly appropriate for situations where reliable subjective weightings cannot be obtained (Deng et al., 2000). These approaches can be classified as follows.

5.5.2.1 Mean weighting

The Mean Weighting (MW) method (Deng et al., 2000) determines objective weightings by $w_j = \frac{1}{n}$, where n is the number of criteria. This is based on the assumption that all of the attributes are of equal importance. The MW (equal importance) should be used either when there is no information from the DM or when there is not enough information to distinguish the relative importance of criteria.

5.5.2.2 Entropy

According to "Information Theory," entropy is a criterion for the amount of uncertainty represented by a discrete probability distribution in which there is an agreement that a broad distribution represents more uncertainty than does a sharply packed one (Pratyyush and Jian-Bo, 1998). In the Entropy method (Hwang and Yoon, 1981; Shanian and Savadogo, 2009; Shanian and Savadogo, 2006b), the attributes with performance ratings that are very different from each other, have higher importance for the problem due to more influence on ranking outcomes (Zeleny, 1982; Asgharpour, 1999). In other words, an attribute has less importance if all candidate materials have similar performance ratings for that attribute. The method determines the weightings of the attributes using Eqs. (5.9)−(5.11).

$$p_{ij} = \frac{x_{ij}}{\sum\limits_{i=1}^{m} x_{ij}} \quad i = 1,\ldots,m; \quad j = 1,\ldots,n \tag{5.9}$$

$$E_j = -\left(\sum_{i=1}^{m} p_{ij}\ln(p_{ij})\right)/\ln(m) \quad j = 1,\ldots,n \tag{5.10}$$

$$w_j = \frac{1 - E_j}{\sum_{k=1}^{n}(1 - E_k)} \quad j = 1,\ldots,n \tag{5.11}$$

5.5.2.3   Standard deviation method

The Standard Deviation (SD) method (Diakoulaki et al., 1995) is similar to the Entropy method, in that it assigns a small weighting to an attribute if it has similar attribute values across alternatives. The SD method determines the weightings of the criteria in terms of their SDs using Eqs. (5.12) and (5.13). The application of this method in materials selection was recently suggested by Rao and Patel (2010).

$$w_j = \frac{\sigma_j}{\sum_{j=1}^{n} \sigma_j} \quad j = 1,\ldots,n \tag{5.12}$$

$$\sigma_j = \sqrt{\frac{\sum_{i=1}^{m}(x_{ij} - \bar{x}_j)^2}{m}} \quad j = 1,\ldots,n \tag{5.13}$$

where $\bar{x}_j$ is the average of the data in the considered criterion.

The SD method is not as accurate as the Entropy method because its results may be affected by the range of different criteria, while the normalization process (Eq. (5.9)) in entropy prevents this from being misleading.

5.5.2.4   Criteria importance through intercriteria correlation

MCDM involves determining the optimal alternative among multiple, conflicting, and interactive criteria (Chen et al., 1992). In MCDM, many of the criteria are often highly correlated (Ramík and Perzina, 2010; Angilella et al., 2004), and the incorporation of several interdependent criteria could yield misleading results, while the arbitrary omission of some criteria entails the removal of more or less useful information sources (Diakoulaki et al., 1995). Furthermore, an attribute cannot often be considered separately because of the complementarities between them. For example, in the case of steel, there is a common relationship between the Brinell hardness number (BHN) and the Ultimate Tensile Strength (UTS). Similar relationships can be shown for brass, aluminum, and cast irons. These kinds of relationships have been reported widely in materials engineering for different mechanical properties (Durst et al., 2008; Jiang et al., 2006). Moreover, in the conceptual design stage, when designers are more interested in sensorial aspects of materials (Karana et al., 2008; Karana et al., 2009), the interdependency would be more significant because the technical and sensorial properties of materials have to be considered simultaneously and these two have an obvious relationship. For instance, both sensorial criteria of transparency and smoothness are used for conveying the meaning of sexy in a product (Karana et al., 2009), while there are relationships between these two aspects and mechanical properties. Considering these interdependencies may reduce the risk of the wrong selection when there are a lot of materials with very similar performances. One way to address this issue is to obtain the relationship among criteria and then to derive the final weightings by considering the influences between them. An objective weighting method of Criteria Importance Through Intercriteria Correlation (CRITIC) based on the SD method was proposed by Diakoulaki et al. (1995).

5.5.3 Integrated methods

Sometimes the weightings determined by objective methods are inconsistent with the DM's subjective preferences. Conversely, the judgments of the DMs occasionally depend on their knowledge or experience, and the error in weightings to some extent is unavoidable. It can be seen that none of the two approaches are perfect and an integrated method might be the most appropriate for determining the criteria weightings. Currently, a number of combinations (Wang and Parkan, 2006; Jee and Kang, 2000) or optimal weighting methods have been proposed and developed by researchers. Ma et al. (1999) proposed a subjective and objective integrated approach to determine attribute weightings using a mathematical programming model. Xu (2004) showed that the objective weighting method introduced by Ma et al. (1999) does not adhere to the rule of the entropy method. Furthermore, recently Wang and Luo (2010) proposed a method for determining the weightings of attributes based on integration of Correlation Coefficient (CC) and SD such that the DM's subjective preferences can be taken into account. However, this method needs a lot of computation and does not consider pair-wise comparison of attributes. Also, it was revealed that CC and SD weightings may sometimes be close to CRITIC weightings.

Recently, a combinative weighting method was presented by Jahan et al. (2012b) for objective, subjective and correlation weightings that are able to strengthen the existing MCDM materials selection procedures, especially when there are numerous alternatives with interrelated criteria. In this research, it was shown that the traditional assumption used in MCDM modeling that the criteria should be independent has not been established in materials selection. Ignoring the dependency of criteria, makes the model unrealistic and the DM who accepts an optimal solution from the model cannot be sure that he/she has made the correct trade-offs among the objectives. Thus, a systematic framework for weighting was developed to provide insights for designers and other DMs, and to overcome the shortcomings of the current methods. An advantage of the method over the classical approaches is that it does not require the hypothesis of preferential independence; and it may be considered more comfortable and appropriate in materials selection process. The method for incorporating all kinds of weightings can help to avoid the subjectivity from the personal bias of the designer and confirm the objectivity, so it provides a procedure to acquire stronger decisions. Materials selection by allowing "feedback" and "interactions" within and between sets of design criteria and alternatives has recently been raised by Milani et al. (2012) using the Analytic Network Process (ANP). Although ANP (Milani et al., 2012) can capture this matter, it needs more comparisons than the AHP and it might be demanding. Furthermore, the key for the ANP is to determine the relationship structure among features in advance (Lee and Kim, 2000) and to answer the questions precisely, while it is usually hard for DM to give the true relationship structure by considering many criteria (especially in materials selection). Moreover, according to Jee and Kang (2000), the procedure of materials selection should be objective in order to minimize personal bias and time for developing a new product design.

Due to the importance of applying the correct trade-offs among the objectives, a primary procedure which covers objective, subjective, and dependency weightings (Jahan et al., 2012b) is explained in the following steps:

Step (1): Normalizing the decision matrix using Eq. (5.14)

$$r_{ij} = 1 - \frac{|x_{ij} - T_j|}{\text{Max}\{x_j^{\max}, T_j\} - \text{Min}\{x_j^{\min}, T_j\}} \tag{5.14}$$

where x_{ij} is the rating of alternative i (A_i or material $i = 1, \ldots, m$) with respect to criterion j (c_j or material properties $j = 1, \ldots, n$) in decision matrix. T_j is either the most favorable element (x_{ij}) or the target value in criteria j, $x_j^{\max}$ is maximum element in criterion j, $x_j^{\min}$ is minimum element in criterion j.

Step (2): Calculation of objective weighting (w_j^o) for situations in which either all data are quantitative, or the qualitative data are convertible to the corresponding numbers, via Eqs. (5.15) and (5.16), where $\bar{r}_j$ is the average of normalized values in criterion j (Eq. (5.17)); otherwise use only subjective weighting (w_j^s) (Fig. 5.7). This step has been updated because the Entropy method (Jahan et al., 2012b) cannot cover target criteria. Moreover, applying the SD method after normalizing the data (based on Eq. (5.14)) not only led to cover target criteria, but also eliminates misleading effects of different scales in material properties (Jahan et al., 2012a).

$$\sigma_j = \sqrt{\frac{\sum_{i=1}^{m} (r_{ij} - \bar{r}_j)^2}{m}} \quad j = 1, \ldots, n \tag{5.15}$$

$$w_j^o = \frac{\sigma_j}{\sum_{j=1}^{n} \sigma_j} \quad j = 1, \ldots, n \tag{5.16}$$

$$\bar{r}_j = \frac{\sum_{i=1}^{m} r_{ij}}{m} \quad j = 1, \ldots, n \tag{5.17}$$

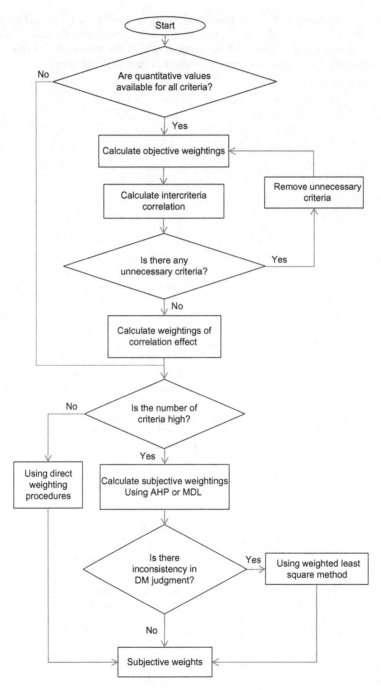

Figure 5.7 Flowchart for objective, subjective, and dependency weighting in materials selection.

Step (3): Calculation of the intercriteria correlation (w_j^c) using Eq. (5.18) and correlation of weighting by Eq. (5.19). Eq. (5.18) uses normalized data (with the same objectives) instead of original data (with different objectives). If the number of criteria is high, decision-making to remove unnecessary criteria can be made at this stage.

$$R_{jk} = \frac{\sum_{i=1}^{m}(r_{ij} - \bar{r}_j)(r_{ik} - \bar{r}_k)}{\sqrt{\sum_{i=1}^{m}(r_{ij}-\bar{r}_j)^2 \sum_{i=1}^{m}(r_{ik}-\bar{r}_k)^2}} \tag{5.18}$$

$$w_j^c = \frac{\sum_{k=1}^{n}(1 - R_{jk})}{\sum_{j=1}^{n}\left(\sum_{k=1}^{n}(1 - R_{jk})\right)} \quad j = 1, 2, 3, \ldots, n \tag{5.19}$$

In Eq. (18), m is the number of materials, n is the number of criteria, $\bar{x}_j$ and $\bar{x}_k$ are the average values of criteria j and k, and R_{jk} is the correlation between criteria j and k. A value of R near 0 indicates little correlation between criteria, while a value near 1 or -1 indicates a high level of correlation. An excessive set of criteria leads to more analytical effort and can make communication with the results of the analysis more difficult. Yurdakul and Tansel (2009) suggested limiting the number of the criteria to around seven, because models with a lower number of criteria are usually more sensitive to changes in weightings of criteria. Any decision-making to remove a criterion from the decision matrix should be carried out carefully based on the idea of DM. Moreover, according to Fig. 5.7, high correlation with a criterion or other criteria needs to be considered as well as less objective weightings.

Step (4): According to Fig. 5.7, in the situation with the low number of criteria, direct weighting techniques are suggested for subjective weighting and, either MDL or AHP is proposed for the high number of criteria. The weighted Least Square method is also suggested when there is inconsistency in DM judgments because it is easier than the eigenvector approach (Chu et al., 1979). In materials selection, researchers always look for logical and simple methods to help designers and DMs in engineering design applications.

Step (5) Incorporating the weightings. Eq. (5.20) combines the three types of weightings.

$$W_j = w_j^s \lambda + w_j^o \frac{(1 - \lambda)}{2} + w_j^c \frac{(1 - \lambda)}{2} \quad j = 1, 2, 3, \ldots, n \tag{5.20}$$

where w_j^o, w_j^s, and w_j^c are the objective, subjective, and correlation effects weighting respectively, and $0 \le \lambda \le 1$.

The formula above enables designers to determine how much importance they wish to assign objective, subjective, and correlation effects on weightings. It is the main advantage of the new method to the current combination technique (Jahan et al., 2012b). In the equation above, $\sum_{j=1}^{n} W_j$ would be equal 1, because $\sum_{j=1}^{n} w_j^s = 1$, $\sum_{j=1}^{n} w_j^o = 1$, $\sum_{j=1}^{n} w_j^c = 1$ and $\lambda + \frac{(1 - \lambda)}{2} + \frac{(1 - \lambda)}{2} = 1$.

Further, by changing the λ between 0 and 1, the suggested formula provides an opportunity for sensitivity analysis of weightings in decision-making problems. Table 5.13 shows various methods for combining the weightings when there are uncertainty in importance of each type of weightings and when all types of weightings have the same importance from the designer's perspective.

Table 5.13 Different methods for combining weightings

	Uncertainty in importance of each type of weightings	Equal importance on each type of weightings
Two types of weightings	$W_j = w_j^1 \lambda + w_j^2 (1 - \lambda)$	$W_j = \dfrac{(w_j^1 * w_j^2)^{\frac{1}{2}}}{\sum\limits_{j=1}^{n} (w_j^1 * w_j^2)^{\frac{1}{2}}}$
Three types of weightings	$W_j = w_j^s \lambda + w_j^o \dfrac{(1 - \lambda)}{2} + w_j^c \dfrac{(1 - \lambda)}{2}$	$W_j = \dfrac{(w_j^s * w_j^o * w_j^c)^{\frac{1}{3}}}{\sum\limits_{j=1}^{n} (w_j^s * w_j^o * w_j^c)^{\frac{1}{3}}}$

In the integration of weightings, the coefficient of λ assigns to subjective weighting and the rest until one $(1 - \lambda)$ assign equally to objective and dependency weightings. This procedure helps to retain the importance of subjective weighting in the combination of weightings. Although λ can vary from 0 to 1, the coefficients in the integration formula (Eq. (5.20)) do not change symmetrically by altering the λ. As a consequence, in the sensitivity analysis of weighting to the ranking of alternatives, low values of λ might be useful for inexperienced designers designing new products where past experience is either not enough or unavailable about the design requirements and long-term effects of material properties. In contrast, high values of λ can be appropriate for those who have adequate knowledge, that is, more experienced designers, and are approximately sure about the subjective weighting. The proposed mathematical model enables designers to calculate the three types of weightings used for cost, benefit, and target criteria in materials selection. Also, it enables them to incorporate different weightings under uncertainty.

In this framework, objective weights are prerequisites of subjective weights. Also it is able to reduce the number of criteria systematically. Based on the result of the correlation test (Yurdakul and Tansel, 2009), if it can be concluded that there is a relationship between two criteria, one of them will be considered adequate and therefore the other one can be eliminated. The idea of the correlation's effect on the weighting is originally about this issue, that is, when correlation of criteria with other attributes is high, it should have less importance due to the role of the other criteria.

5.6 Some recent MADM methods being used in materials selection

Most of the methods proposed for ranking in materials selection have tended to focus on cost and benefit criteria, with target values receiving much less attention. This is in spite of their importance in many practical decision-making problems such as selecting materials to best match the properties of human tissue in biomedical engineering applications. Empowering materials databases with decision-making tools are of significance because the available set of materials is rapidly growing both in type and number (Roth et al., 1994). Among the many methods applied to materials selection (Jahan et al., 2010), it seems that the TOPSIS and VIKOR models from the different MADM techniques are suitable for linking to spreadsheet software to help address the limitation of current computer-aided materials selection systems (Jahan et al., 2010) and enhance efficiency of digital tools (Ramalhete et al., 2010) in materials selection.

Furthermore, the methods are not only able to address fuzzy data in materials selection but also the extended version of the techniques (Jahan et al., 2011b; Jahan et al., 2012a) can cover target criteria in materials selection as well as cost and benefit. A spreadsheet can be created based on these approaches in minutes and any number of materials can be evaluated. An advantage of the spreadsheet approach is its simplicity. It can also take into account any number of material properties and does not involve difficult arithmetic techniques or expensive software. The output of the method also gives numerical values and this allows a ranked shortlist to be created and also means that the suitability of each material can be directly compared.

5.6.1 Extended model of TOPSIS according to target-based normalization technique

The TOPSIS technique is based on the principle that the optimal point should have the shortest distance from the Positive Ideal Solution (PIS) and the farthest from the Negative Ideal Solution (NIS). Therefore, this method is suitable for risk avoidance designer(s), because the designer(s) might like to have a decision which not only makes as much profit as possible, but also avoids as much risk as possible (Opricovic and Tzeng, 2004). TOPSIS has been used predominantly in materials selection due to its superior characteristics (Jahan et al., 2010). The extended model of TOPSIS (Jahan et al., 2012a), which has been developed for the selection of materials based on target values of criteria, is explained in this section. The steps of the method are as follows:

1. Convert the raw measures x_{ij} into the standardized measures r_{ij} according to the proposed normalization technique in Eq. (5.21).

$$r_{ij} = 1 - \frac{|x_{ij} - T_j|}{\text{Max}\{x_{ij}^{\text{max}}, T_j\} - \text{Min}\{x_{ij}^{\text{min}}, T_j\}} \tag{5.21}$$

where T_j is either the most favorable element (x_{ij}) or the target value in criteria j.

2. Develop a set of importance weightings (w_j) for criteria.
3. Multiply the columns of the normalized decision matrix by the associated weightings (Eq. (5.22)).

$$V_{ij} = r_{ij}w_j; \quad j = 1, 2, 3, \ldots, n; \quad i = 1, 2, 3, \ldots, m \tag{5.22}$$

4. Identify the PIS.

$$\{V_1^+, V_2^+, V_3^+, \ldots, V_n^+\} = \{\max_i V_{ij} | i = 1, \ldots, m\}$$

5. Identify the NIS.

$$\{V_1^-, V_2^-, V_3^-, \ldots, V_n^-\} = \{\min_i V_{ij} | i = 1, \ldots, m\}$$

6. Develop a distance measure for each alternative to both ideal (D^+) and nadir (D^-) using Eqs. (5.23) and (5.24).

$$D_i^+ = \left(\sum_{j=1}^{n} (V_{ij} - V_j^+)^2\right)^{0.5} \quad i = 1, 2, 3, \ldots m \tag{5.23}$$

$$D_i^- = \left(\sum_{j=1}^{n} (V_{ij} - V_j^-)^2\right)^{0.5} \quad i = 1, 2, 3, \ldots m \tag{5.24}$$

7. Calculate the relative closeness to the ideal solution according to Eq. (5.25).

$$C_i = \frac{D_i^-}{D_i^- + D_i^+}; \quad i = 1, 2, 3, \ldots m; \quad 0 < C_i < 1 \tag{5.25}$$

8. Rank alternatives by maximizing the ratio in Step 7. The larger the index value, the better the performance of the alternative.

Steps 4 and 5 in the extended TOPSIS are simplified compared to the original TOPSIS because of the advantages of the new normalization procedure compared to the vector normalization approach that is used in the original TOPSIS.

5.6.2 Comprehensive VIKOR

The VIKOR method was developed for multi-criteria optimization in complex systems (Opricovic and Tzeng, 2004) and enjoys wide acceptance. It focuses on ranking and selecting from the alternatives with conflicting and different units criteria. In the VIKOR approach, the compromise ranking is performed by comparing the measure of closeness to the ideal alternative, and compromise achieved through agreement established by mutual concessions. Chang (2010) developed the modified VIKOR method to avoid numerical difficulties in solving problems by the traditional VIKOR method. In this section, the modified VIKOR is adapted using a novel normalization technique. The advantage of proposing a comprehensive and compromising model to the traditional VIKOR is that it covers all objectives in MCDM. Furthermore, the suggested model overcomes the critical problem of VIKOR, which was demonstrated by Huang et al. (2009). In the suggested model, the qualitative attributes (linguistic terms) can be systematically converted to their

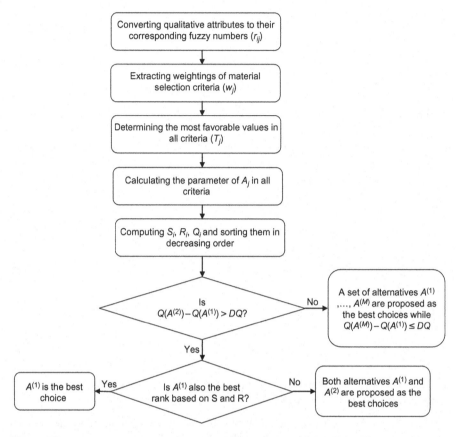

Figure 5.8 Flowchart of comprehensive VIKOR method for materials selection.

corresponding fuzzy numbers using eleven-point conversion scales (Rao, 2006, 2008; Rao and Davim, 2008; Rao and Patel, 2010) in materials selection. The comprehensive VIKOR method (Jahan et al., 2011b) covers all types of criteria and overcomes the main error of traditional VIKOR by using a simpler approach. A flowchart of the comprehensive version of VIKOR method is presented in Fig. 5.8, with details of the method illustrated step by step as follows.

Step 1: Determine the most favorable values for all criteria.

$$T = \{T_1, T_2, T_3, \ldots, T_j, \ldots, T_n\} = \{\text{Most desirable element } (x_{ij}) \text{ or target value for criteria } j\}.$$

where $x(i = 1, 2, 3, \ldots, m$ and $j = 1, 2, 3, \ldots, n)$ are elements of the decision matrix (alternative i respect to criteria j).

Step 2: Compute the values S_i and R_i by Eqs. (5.26) and (5.27).

$$S_i = \sum_{j=1}^{n} w_j \left(1 - e^{\frac{|x_{ij} - T_j|}{-A_j}} \right) \qquad (5.26)$$

$$R_i = \max_j \left[w_j \left(1 - e^{\frac{|x_{ij} - T_j|}{-A_j}} \right) \right] \qquad (5.27)$$

where $A_j = \begin{cases} 1 & \text{if elements of criteria } j \text{ are normalized between 0 and 1} \\ \max\{x_j^{\max}, T_j\} - \min\{x_j^{\min}, T_j\} & \text{otherwise} \end{cases}$, $x_j^{\max}$ and $x_j^{\min}$ are maximum and minimum elements in criteria j respectively and w_j represents the weightings of criteria j (relative importance).

Step 3: Compute the index values Q_i according to Eq. (5.28). These index values are defined as:

$$Q_i = \begin{cases} \left[\dfrac{R_i - R^-}{R^+ - R^-} \right] & \text{if } S^+ = S^- \\[3ex] \left[\dfrac{S_i - S^-}{S^+ - S^-} \right] & \text{if } R^+ = R^- \\[3ex] \left[\dfrac{S_i - S^-}{S^+ - S^-} \right] v + \left[\dfrac{R_i - R^-}{R^+ - R^-} \right] (1 - v) & \text{otherwise} \end{cases} \qquad (5.28)$$

where $S^- = \min S_i$, $S^+ = \max S_i$, $R^- = \min R_i$, $R^+ = \max R_i$, and v is introduced as a weighting for the strategy of "the majority of criteria" (or "the maximum group utility"), whereas $1 - v$ is the weighting of the individual regret. The value of v lies in the range of $0-1$ and these strategies can be compromised by $v = 0.5$.

Step 4: The results are three ranking lists. By sorting the values S, R, and Q in decreasing order.

Step 5: Propose, as a compromise solution, the alternative $(A^{(1)})$, which is the best ranked by the measure Q (minimum) if the following two conditions are satisfied:

    C1. Acceptable advantage:

$$Q(A^{(2)}) - Q(A^{(1)}) \geq DQ$$

Where $A^{(2)}$ is the alternative with second place in the ranking list by Q; $DQ = 1/(M - 1)$. M is the number of alternatives.

    C2. Acceptable stability in decision-making:

        The alternative $A^{(1)}$ should also be the best ranked by S or/and R. A set of compromise solutions is proposed as follow, if one of the conditions is not satisfied.

- Alternatives $A^{(1)}$ and $A^{(2)}$ if only the C2 is not satisfied, or
- Alternatives $A^{(1)}$, $A^{(2)}$, ..., $A^{(M)}$ if the C1 is not satisfied; $A^{(M)}$ is determined by the relation $Q(A^{(M)}) - Q(A^{(1)}) < DQ$ for maximum M.

It also appears that the comprehensive VIKOR and extended TOPSIS that cover target criteria are suitable for linking to materials databases as a means of enhancing the efficiency of computer-based materials selection tools. It is clear that decision-making techniques that address target criteria as well as cost and benefit criteria can help engineering designers make better informed choices of materials. From a design perspective, the improved accuracy of the final ranking of alternative materials using the techniques described imparts confidence in decision-making and hence reduces the incidences of costly mistakes.

5.6.3 An introduction to fuzzy MADM methods for materials selection

One of the most important aspects of solving MCDM problems is the precise evaluation of information, particularly in the quantitative form. For evaluation, and judgment of such data, natural language is often employed in order to articulate thinking and subjective perceptions. Most of the design decision-making problems, quite often involve imprecise or fuzzy information about the criteria and alternatives, while traditional MCDM methods fail to handle the imprecision and uncertainty of information. Thus fuzzy MCDM techniques have evolved as possible solutions to these complex situations. However, MCDM problems may become very complex when the score of alternatives with respect to criteria are difficult to be quantified precisely (Majumdar, 2010).

For material properties, such as corrosion and wear resistance, machinability, and weldability, numerical values are rarely given and materials are usually rated as very good, good, fair, poor, etc. Material properties are also of varying degrees of significance for different design requirements. Therefore, the score of materials and importance weighting of design selection criteria are usually described in a linguistic fashion. In such cases, fuzzy logic can be very useful. Fuzzy set theory was developed exactly based on the idea that the key elements in human thinking are not numbers, but linguistic terms or labels of fuzzy sets. The application of fuzzy logic in materials and process selection would be useful, because these decisions are made during the preliminary design stages, that is, in a situation characterized by imprecise and uncertain requirements, parameters, and relationships (Giachetti, 1998). Using fuzzy analysis in the earliest stages of preliminary design evaluation or in situations limited to semantic input from design DMs was recommended (Thurston and Carnahan, 1992). Also, in engineering design, the choice between either design concepts or materials, is usually a team-based effort, involving designers, clients, and managers, who all may have different opinions on what is the "best." Another important advantage of fuzzy multiple criteria methods is their capability of addressing the problems that are marked by different conflicting interests (Mardani et al., 2015), or group decision-making. The combination of all these various single preferences into a final decision that selects the "best" design among various alternatives needs a systematic method. Fig. 5.9 demonstrates the capabilities of fuzzy MADM methods for materials selection problems. As described, the fuzzy methods can take into account the view points of the design team in materials selection process for both qualitative and incomplete data, in addition to quantitative data. Group and individual direct

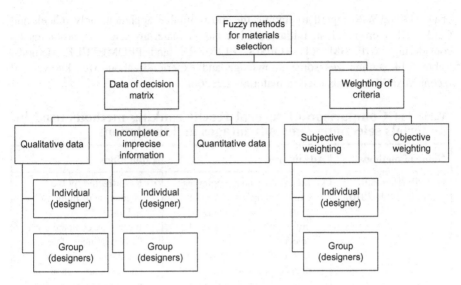

Figure 5.9 Capabilities of fuzzy MADM methods for materials selection problems.

weighting methods are also useable in fuzzy MADM materials selection problems. Since the inputs provided by the DM is in a linguistic form, there is a possible chance of getting incomplete problems (Jeya Girubha and Vinodh, 2012).

For a better understanding, a materials selection example using fuzzy methods for ranking of materials is described in Chapter 7.

5.7 Aggregation method for complex materials ranking problems

MADM methods differ in criteria weightings determination, normalization, and aggregation; as a result different MADM methods regularly create different outcomes for selecting or ranking a set of alternative decisions involving multi-criteria (Yeh, 2002). Voogd (1983) showed that for at least 40% of the time each technique produced a different result from any other techniques. This leads to two highly important questions for which both are difficult to answer, "how can a consensus ranking be found when several different MADM methods rank differently?" and "which technique is the most suitable to solve the problem?" Considerable effort has been spent on the development of numerous MADM models but despite all of these there is no best approach and no single multi-criteria analysis technique that is inherently better than the others (Hajkowicz and Collins, 2007). These numerous existing methods (Jahan et al., 2010; Chatterjee et al., 2011) do not simplify the decision-making problem but make it relatively complicated. A decision on the most appropriate decision-making approach may be viewed as a difficult problem for DMs, hence the selection of the method itself is a complex MADM problem (Li et al., 2008). With regard to finding the most suitable approach, only (Cicek and Celik, 2010) proposed an initial decision aid in materials selection problems by considering AHP, SMART, TOPSIS, ELECTRE, and PROMETHEE methods. Table 5.14 summarizes some knowledge and experience about well-known and recent MADM methods used in materials selection.

Table 5.14 Comparison of several recent ranking methods used in materials selection: some advantages and limitations

Name of method	Advantages	Limitations
• AHP (Hambali et al., 2011, 2010)	• It can combine the opinions expressed by several groups of experts to attain either criteria weightings or final selection. • It is a powerful and flexible decision-making procedure especially when both tangible and nontangible criteria are available.	• It is affected by the vagueness of human judgments. • It can only compare a very limited number of decision alternatives. When there are hundreds or thousands of options to be compared, the pair wise comparison is infeasible.
• ANP (Milani et al., 2012)	• It is an extended version of AHP and considers feedback and interaction between decision factors and/or alternatives.	• It is affected by the vagueness of human judgments and needs more comparisons than the AHP.
• SAW method (Dehghan-Manshadi et al., 2007; Fayazbakhsh et al., 2009)	• It is very simple.	• It is not very sensitive to the values of criteria compared to TOPSIS (Simanaviciene and Ustinovichius, 2010).
• Graph theory and matrix approach (2006)	• It can consider any number of quantitative and qualitative factors.	• It does not have a condition for checking the consistency made in the judgments of relative importance of the attributes. • The method may be difficult to deal with if the number of attributes is more than 20.
• ELECTRE (Chatterjee et al., 2009, Shanian and Savadogo, 2006a, 2006c; Milani and Shanian, 2006)	• It is more appropriate when the numbers of criteria are more than the alternatives. • An important advantage of this technique is that indifference and preference	• As the number of alternatives increases, the amount of calculations increases quite rapidly. • It can only determine the rank of each material and

(Continued)

Table 5.14 (Continued)

Name of method	Advantages	Limitations
• Original version of TOPSIS	thresholds can be considered when modeling the imperfect knowledge of data. • It is able to evaluate purely ordinal scales, without the need to convert the original scales (Figueira et al., 2009). • It is useful when dealing with a large number of alternatives and criteria, good for both qualitative and quantitative data, relatively easy and fast. • The output can be a preferential ranking of the candidate materials with a numerical value that provides a better understanding of differences and similarities among alternatives.	does not give numerical values for better understanding of differences between alternatives. • It is suitable for risk avoidance designers (Opricovic and Tzeng, 2004).
• Original version of VIKOR (Chatterjee et al., 2009, Rao, 2008)	• It has all the advantages of TOPSIS.	• It is suitable for those situations in which the designer wants to obtain maximum yield.

Furthermore, the outcome in inconsistency of MADM methods increases either as the number of alternatives increases (Olson et al., 1995), or the alternatives have similar performance (Olson et al., 1995; Shanian and Savadogo, 2009). If all the alternatives' ranking orders in different MADM methods are similar, the decision-making process will be ended. Otherwise, the ranking results cannot be valid and reliable. If the ranking outcomes from different techniques differ significantly, the validity issue will be crucial (Hobbs et al., 1992).

The application of various MADM methods can yield different results, especially when alternatives lead to similar performance. Therefore, an aggregation technique is proposed for optimal decision-making.

Some researchers have suggested applying different MADM methods together as a more efficient design tool in order to enhance the accuracy of the final decision and for the sake of making safer engineering decisions when the difference between the alternative solutions are inherently close together (Liao, 1996; Ashby and Johnson, 2010a; Jahan et al., 2012a). The aggregation of individual rankings by various MADM methods is usually done by an averaging function (Ataei, 2010) as a basic aggregation strategy. However, when using this process, there is no guarantee of obtaining optimum results for circumstances in which there are large differences between the rankings of alternatives. As a consequence, Borda and Copeland rules (Pomerol and Barba-Romero, 2000), the most common voting aggregation techniques in group decision-making (Hwang and Lin, 1987) are used for aggregation of MADM results (Ataei, 2010). The Borda rule assigns more points to higher rankings and then adds up the points over all individual voters for every alternative. The option that has the highest points in the voters' rankings is then chosen. The Copeland Rule is a single-winner strategy in which the winner is identified by finding the candidate with the most pair-wise victories. Favardin et al. (2002) showed that the Borda rule is significantly more vulnerable than the Copeland Rule, although the probability of a tied situation is the main weakness of both techniques (Hwang and Lin, 1987).

Therefore, the basis of an integrated framework for assisting complex decision-making in materials selection is described. The key stages of an aggregation method (Jahan et al., 2011a) are described step by step. The input is the ranking orders of materials by different methods. The main steps of the model are as follows:

Step 1: Generate a material-ranking matrix as a $(m \times m)$ square matrix in which M_{ik} shows the number of times each candidate material i assigns to the kth ranking. In other words, M_{ik} measures the contribution of material i to kth ranking.

Step 2: Calculate C_{ik}, which is $M_{ik} + C_{i,k-1}$ $i, k = 1, \ldots, m C_{i,0} = 0$. This step helps to achieve optimal assignment of the materials to the different rankings.

Step 3: Obtain the most suitable material for each ranking. A Linear Programming (LP) model is suggested for assigning the problem. Due to the importance of high rankings, the objective function can be weighted by $W_k = \frac{m^2}{k}$ (where m is the number of materials), which generates more weighting for top rankings. Now, a permutation square matrix N as $(m \times m)$ is defined in which element $N_{ik} = 1$ if the kth ranking assign to alternative i and $N_{ik} = 0$ otherwise. The aggregation model can be written in the following LP format:

$$\text{Max} \sum_{i=1}^{m} \sum_{k=1}^{m} C_{ik} \times W_k \times N_{ik} \tag{5.29}$$

$$\sum_{k=1}^{m} N_{ik} = 1 \quad i = 1, 2, 3, \ldots, m \tag{5.30}$$

$$\sum_{i=1}^{m} N_{ik} = 1 \quad k = 1, 2, 3, \ldots, m \tag{5.31}$$

$$N_{ik} = \begin{cases} 0 \\ 1 \end{cases} \quad \text{for all } i \text{ and } k$$

Eq. (5.29) is the objective function, which attempts to give higher weighting for top rankings and tries to assign each ranking to the suitable material. Eq. (5.30) indicates that each

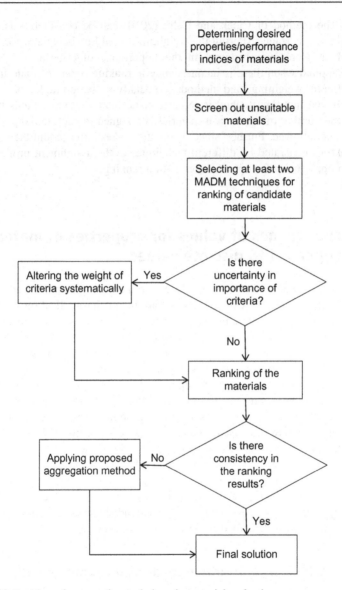

Figure 5.10 Position of aggregation technique in materials selection process.

alternative has only one ranking, and Eq. (5.31) determines that each ranking is assigned to only one alternative. The regular simplex algorithm can be used to solve the above LP problem.

The position of the proposed aggregation procedure in the materials selection process is shown in Fig. 5.10. After translating the design requirements into material requirements and screening out unsuitable materials, the process starts by choosing some independent and appropriate MADM methods to rank the alternative materials. The method of Cicek and Celik (2010) can be considered as an initial decision aid at this stage. Altering the weightings would be beneficial for the situation in which there is uncertainty in the importance of criteria. The suggested method is applied when there is inconsistency in ranking orders of materials generated by different weightings and methods, especially in the top rankings.

It is believed that this method is more realistic than current methods because it considers the number of times each material is assigned to each ranking, and a tied situation is not allowed. Further, since the method takes into account the importance of top rankings generated by different techniques in the assignment process, it will provide an opportunity for optimal aggregation ranking.

5.8 Use of ranges of values for properties of materials as opposed to discrete values

The properties of materials are normally determined using standardized test methods. This ensures both consistency and reliability in the application of materials. However, there is always variability in measurement systems and in manufacturing processes, therefore material properties often vary to some degree (Ashby and Johnson, 2010b). The properties of materials often behave linearly in a given operating range, and large factors of safety will be used in design calculations when that range is modeled as a constant. Although using either the middle value or average value can significantly simplify design calculations and therefore materials selection, this simplification often leads to missing some information, which is very important in the failure of resistant materials (e.g., UTS of material).

The interval numbers are the simplest form of representing uncertainty in the decision matrix. Table 5.15 shows a typical MADM problem with interval data, where x_{ij}^L and x_{ij}^U are the rating of alternative i (A_i or material i, $i = 1,\ldots,m$) with respect to criterion j (c_j or material property j, $j = 1,\ldots,n$) in decision matrix. The interval target-based VIKOR, which is the extended version of interval VIKOR (Sayadi et al., 2009) is described step by step as follows.

Table 5.15 A typical multi-attribute decision problem with interval data

	w_1 C_1	w_2 C_2	$\ldots$	w_n C_n
A_1	$[x_{11}^L, x_{11}^U]$	$[x_{12}^L, x_{12}^U]$	$\ldots$	$[x_{1n}^L, x_{1n}^U]$
A_2	$[x_{21}^L, x_{21}^U]$	$[x_{22}^L, x_{22}^U]$	$\ldots$	$[x_{2n}^L, x_{2n}^U]$
A_3	$[x_{31}^L, x_{31}^U]$	$[x_{32}^L, x_{32}^U]$	$\ldots$	$[x_{3n}^L, x_{3n}^U]$
$\vdots$	$\vdots$	$\vdots$		$\vdots$
A_m	$[x_{m1}^L, x_{m1}^U]$	$[x_{m2}^L, x_{m2}^U]$	$\ldots$	$[x_{mn}^L, x_{mn}^U]$

Step 1: Determine the most favorable values for all criteria.

$T = \{T_1, T_2, T_3, \ldots, T_j, \ldots, T_n\}$ = {most desirable element or target value for criteria j}.

Step 2: Compute the values S_i^L, S_i^U, R_i^L and R_i^U using Eqs. (5.32)–(5.37).

$$V_{ij}^L = \frac{\left|x_{ij}^L - T_j\right|}{\max\{x_j^{U\max}, T_j\} - \min\{x_j^{L\min}, T_j\}} \tag{5.32}$$

$$V_{ij}^U = \frac{\left|x_{ij}^U - T_j\right|}{\max\{x_j^{U\max}, T_j\} - \min\{x_j^{L\min}, T_j\}} \tag{5.33}$$

$$S_i^L = \sum_{j=1}^{n} w_j \min(V_{ij}^L, V_{ij}^U) \tag{5.34}$$

$$R_i^L = \max_j\{\min(V_{ij}^L, V_{ij}^U)\} \tag{5.35}$$

$$S_i^U = \sum_{j=1}^{n} w_j \max(V_{ij}^L, V_{ij}^U) \tag{5.36}$$

$$R_i^U = \max_j\{\max(V_{ij}^L, V_{ij}^U)\} \tag{5.37}$$

where $x_j^{U\max}$ is maximum element in upper bound of criteria j, $x_j^{L\min}$ is minimum element in lower bound of criteria j, T_j is either the most favorable element or the target value in criteria j and finally w_j is the weighting or relative importance of criteria.

Step 3: Compute the index values Q_i^L and Q_i^U. These index values are defined by Eqs. (5.38) and (5.39), respectively.

$$Q_i^L = \left[\frac{S_i^L - S^-}{S^+ - S^-}\right]\nu + \left[\frac{R_i^L - R^-}{R^+ - R^-}\right](1 - \nu) \tag{5.38}$$

$$Q_i^U = \left[\frac{S_i^U - S^-}{S^+ - S^-}\right]\nu + \left[\frac{R_i^U - R^-}{R^+ - R^-}\right](1 - \nu) \tag{5.39}$$

where $S^- = \min_i S_i^L$, $S^+ = \max_i S_i^U$, $R^- = \min_i R_i^L$, $R^+ = \max_i R_i^U$, and ν is introduced as a weighting for the strategy of "the majority of criteria" (or "the maximum group utility"), whereas $1 - \nu$ is the weighting of the individual regret. The value of ν lies in the range of 0 to 1 and these strategies can be compromised by $\nu = 0.5$.

Step 4: The alternative that has minimum Q_i ($[Q_i^L, Q_i^U]$) is the best alternative and it is chosen as a compromise solution. Here $Q_i, i = 1, \ldots, m$ are interval numbers and to choose the minimum interval number, they are compared with each other as shown in Table 5.16.

Table 5.16 Comparison of interval numbers

Minimum interval number between $[a^L, a^U]$ and $[b^L, b^U]$		
When $a^U < b^L$	$\Rightarrow$	$[a^L, a^U] < [b^L, b^U]$
When $a^L < b^L < b^U < a^U$; If $\alpha(b^L - a^L) \geq (1-\alpha)(a^U - b^U)$	$\Rightarrow$	$[a^L, a^U] < [b^L, b^U]$
When $a^L < b^L < b^U < a^U$; If $\alpha(b^L - a^L) < (1-\alpha)(a^U - b^U)$	$\Rightarrow$	$[b^L, b^U] < [a^L, a^U]$
When $a^L < b^L < a^U < b^U$; If $\alpha(b^L - a^L) \geq (1-\alpha)(b^U - a^U)$	$\Rightarrow$	$[a^L, a^U] < [b^L, b^U]$
When $a^L < b^L < a^U < b^U$; If $\alpha(b^L - a^L) < (1-\alpha)(b^U - a^U)$	$\Rightarrow$	$[b^L, b^U] < [a^L, a^U]$

The parameter α is the optimism level of the DM ($0 < \alpha \leq 1$). The optimist DM has a greater α value than the pessimistic DM. A rational DM is represented by $\alpha = 0.5$.

There is a considerable choice of different materials, and suppliers of materials, available for designing new products (van Kesteren, 2008). Under these circumstances, which is the norm, DMs are not able to use precise values to express their evaluations but they can still provide approximate ranges (e.g., interval numbers) of evaluations using their own knowledge. For novice designers, or designers who only select materials on a casual basis, the problem of optimal materials selection can be overwhelming and lead to poor decision-making. Current methods and computer packages facilitate materials selection but only support specific aspects of the process. There is therefore increasing consideration being given to using MADM methods in materials selection (Zhao et al., 2012; Lin et al., 2012; Karande and Chakraborty, 2012; Maity et al., 2012), especially because the degree of complexity and range of options are growing continually. Verifying the exact values of the attributes is difficult in materials selection therefore, it is more appropriate to consider them as interval numbers. (Swift et al., 2000) highlighted the importance of understanding the variability associated with the properties of materials, manufacturing tolerances and in service loading conditions for engineering design. The limits of the variation of properties are defined in materials and manufacturing process standards. However, material engineers need to report material properties carefully because the materials produced tend to vary from batch to batch from a particular manufacturer, and from manufacturer to manufacturer. Therefore, understanding the variation is crucial, and dealing with interval data is unavoidable in materials selection. Applying the concepts and operations of interval numbers is helpful when dealing with uncertain information.

If the presented range of materials properties have a normal distribution (Farag, 1997), it is better to consider the middle of the interval data as the mean and continue ranking with available techniques for the exact data (Fig. 5.11). In the mass production of materials, extensive testing can be carried out in order to determine the probabilistic distribution of materials' properties. When the production is in small batches, or a material is new, materials testing may be limited in the number of test repetitions to provide statistical relevance. Thus, the probabilistic distribution of materials properties is usually unknown and a different distribution such as a uniform Weibull or an exponential might be fitted to them. In such a situation, there is no need to consider the middle of interval data range and that means the interval

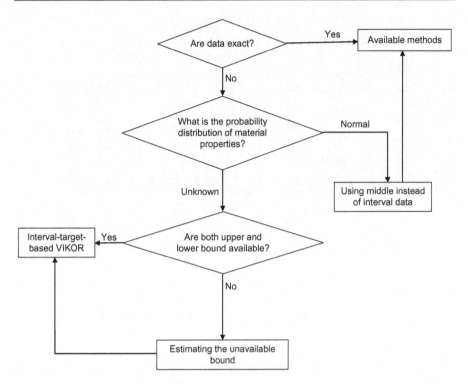

Figure 5.11 Dealing with incomplete and interval data in target-based materials selection problems.

target-based VIKOR method (Jahan and Edwards, 2013) can be used. For incomplete data, where only the upper bound or lower bound is available, it is not possible to continue before estimating the unidentified bound. It is expected that decision models will be developed in this area and more extensive effort and time devoted to decision-making about materials and design.

5.9 Computer implementation – application of spreadsheet and mathematical analysis techniques to facilitate the analysis of ranking problems

From the designers' perspective, the wider utilization of multi-criteria evaluation techniques depends on the availability of powerful and user-friendly oriented software tools. This section describes a simple application for evaluation of alternatives based on extended TOPSIS developed using Microsoft® (MS) Excel spreadsheet software. A template was developed by the authors for describing the structure of the calculations. Since the MADM is in the early stages of development in the area of materials design and selection, using the key functions of MS Excel is

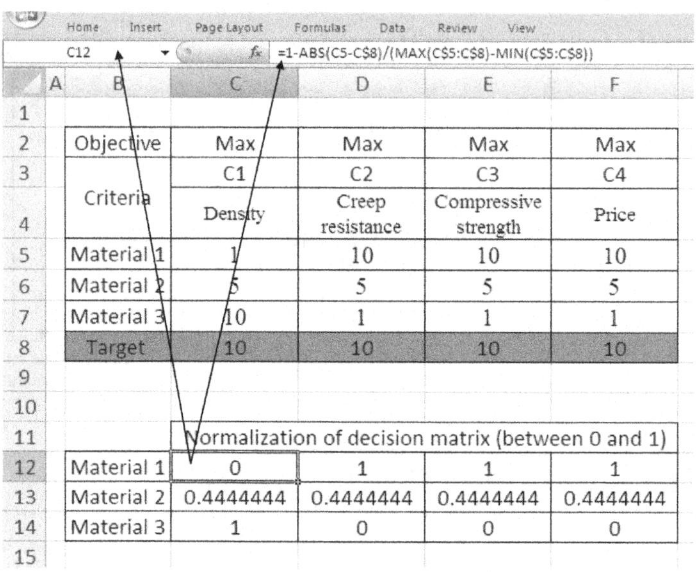

C12 =1-ABS(C5-C$8)/(MAX(C$5:C$8)-MIN(C$5:C$8))

	B	C	D	E	F
2	Objective	Max	Max	Max	Max
3	Criteria	C1	C2	C3	C4
4		Density	Creep resistance	Compressive strength	Price
5	Material 1	1	10	10	10
6	Material 2	5	5	5	5
7	Material 3	10	1	1	1
8	Target	10	10	10	10
11		Normalization of decision matrix (between 0 and 1)			
12	Material 1	0	1	1	1
13	Material 2	0.4444444	0.4444444	0.4444444	0.4444444
14	Material 3	1	0	0	0

Figure 5.12 Decision matrix and formula in MS Excel for normalizing data.

appropriate. Compared to other software products, MS Excel is able to work with interval data as well as capabilities of sensitivity analysis of weightings. This allows creating and managing a decision matrix (add, remove, or insert) with any number of criteria and alternatives, as well as integrating results with charts.

The basic concept of the TOPSIS method is that the selected material options should have the shortest distance from the ideal solution and the farthest from the negative-ideal solution. The Euclidean distance approach is used to evaluate the relative closeness of each material option to the ideal solution. The extended TOPSIS assumes that each criterion has a goal direction (Minimize, Maximize, or Target). Figs. 5.12 and 5.13 show how all the calculations can be stored in a standard MS Excel worksheet.

MMULT (array1, array2), which returns the matrix product of two arrays, and SUMXMY2, which returns the sum of squares of differences of corresponding values in two arrays, are the key required functions in TOPSIS approach. In matrix multiplication, the number of columns in array1 must be the same as the number of rows in array2.

If $A = \begin{bmatrix} -2 & 1 & 3 \\ -4 & 0 & 5 \end{bmatrix}$ and $B = \begin{bmatrix} 2 & 0 \\ 3 & -1 \\ 4 & -3 \end{bmatrix}$, in order to find $E = A \cdot B$, the following steps are required:

a. Enter the matrices A and B anywhere into the Excel sheet (Fig. 5.14). Notice that Matrix A is in cells B4:D5, and Matrix B is in cells G4:H6

b. Multiply Row by Column and the first matrix has 2 rows and the second has 2 columns, so the resulting matrix will have 2 rows by 2 columns. Highlight the cells where the resulting matrix E is to be placed.

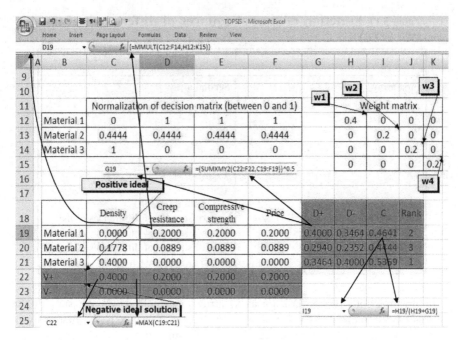

Figure 5.13 Key functions for TOPSIS.

	A	B	C	D	E	F	G	H	I
					{=MMULT(B4:D5,G4:H6)}				
1									
2									
3			A					B	
4		-2	1	3			2	0	
5		-4	0	5			3	-1	
6							4	-3	
7									
8				E=A.B					
9				11	-10				
10				12	-15				
11									

Figure 5.14 Matrix multiplication in MS Excel.

c. Once the resulting matrix is highlighted and while it is still highlighted, enter the following formula: = MMULT(B4:D5,G4:H6).

d. When the formula is entered, press the Ctrl key and the Shift key simultaneously then press the Enter key. This will change the formula just written to: { = MMULT(B4:D5,G4:H6)}. If you do not press these keys simultaneously (holding down Shift and Ctrl then press Enter), the result will appear only in one cell or, an error message will be displayed.

e. The resulting matrix is shown in Fig. 5.14.

Table 5.17 Function of sum of squares of differences of the two arrays

A	B
First array	Second array
2	6
3	5
9	11
1	7
8	5
7	4
5	4
Formula	Description (Result)
= SUMXMY2(A2:A8,B2:B8)	Sum of squares of differences of the two arrays above (79)
= SUMXMY2({2, 3, 9, 1, 8, 7, 5}, {6, 5, 11, 7, 5, 4, 4})	Sum of squares of differences of the two arrays constants (79)

In the function SUMXMY2 (array-x, array-y), which shows the sum of the squares of differences of the two arrays (Table 5.17), array-x is the first array or range of values and array-y is the second array or range of values.

5.10 Summary and conclusions

Matrix format, including alternatives and criteria, are used for materials selection problems. The data for criteria can be exact, fuzzy, incomplete, interval, or stochastic. The selection criteria have different importance, and some of them might have dependency on each other. These criteria can be beneficial (higher is better), nonbeneficial (lower is better), or target-based. The decision-making process might be an individual or a team work activity. Although screening of materials is a prerequisite of the final selection stage, hundreds of materials might remain for the final ranking. All of these elements affect the selection of the appropriate MADM method as well as developing established MADM approaches. The studies that focused on the theoretical aspects of MADM methods in materials selection worked on novel normalization, weighting, or aggregation method. However, there is still a need to enhance the capability of MADM methods based on real or explicit nature of materials selection problems.

Review questions

1. What are the similarities and differences between the TOPSIS and comprehensive VIKOR methods?
2. Classify the weighting methods and explain how each method can be fitted to each condition?

3. What is the weakness of the DL approaches?
4. How should weighting and ranking methods be used when there are dependencies between criteria?
5. What is the advantage of using sensitivity analysis in weighting of criteria?
6. What is the aim of using normalization or dimensionless techniques?
7. Discuss the similarities and differences between target-based normalization methods.
8. What are the shortcomings of the LOP method?
9. What is the effect of material databases on developing specific MADM methods in materials selection?
10. Discuss the advantages of fuzzy MADM methods for materials and design selection problems.
11. Why material properties are sometimes represented by interval data in materials' databases?
12. Why is it necessary to consider interval data in material ranking rather than converting the range in to a single value, and what is the effect of probability of distribution in this regard?
13. When and why do some MADM methods often produce different outcomes for ranking a set of materials and how can the designer find the best material in such situations?
14. Using MS Excel, rank the materials in Table 5.1 by both original and comprehensive VIKOR methods and compare the ranked order using the extended TOPSIS method. How does this affect the choice of material?

References

Angilella, S., Greco, S., Lamantia, F., Matarazzo, B., 2004. Assessing non-additive utility for multicriteria decision aid. Eur. J. Oper. Res. 158, 734—744.

Asgharpour, M.J., 1999. Multiple Criteria Decision Making. Tehran University Publications, Tehran.

Ashby, M., Johnson, K., 2010a. Chapter 1 — Function and personality, Materials and Design. second ed. Butterworth-Heinemann, Oxford.

Ashby, M., Johnson, K., 2010b. Chapter 4 — The stuff ... multi-dimensional materials, Materials and Design. second ed. Butterworth-Heinemann, Oxford.

Ashby, M., Johnson, K., 2010c. Chapter 10 — Conclusions, Materials and Design. second ed. Butterworth-Heinemann, Oxford.

Ashby, M.F., Brechet, Y.J.M., Cebon, D., Salvo, L., 2004. Selection strategies for materials and processes. Mater. Des. 25, 51—67.

Ataei, M., 2010. Multi-Criteria Decision Making. Shahrood University of Technology, Shahrood.

Bahraminasab, M., Jahan, A., 2011. Material selection for femoral component of total knee replacement using comprehensive VIKOR. Mater. Des. 32, 4471—4477.

Bahraminasab, M., Sahari, B.B., Edwards, K.L., Farahmand, F., Jahan, A., Hong, T.S., et al., 2014. On the influence of shape and material used for the femoral component pegs in knee prostheses for reducing the problem of aseptic loosening. Mater. Des. , 416—428.

Bottomley, P.A., Doyle, J.R., 2001. A comparison of three weight elicitation methods: good, better, and best. Omega. 29, 553—560.

Cavallini, C., Giorgetti, A., Citti, P., Nicolaie, F., 2013. Integral aided method for material selection based on quality function deployment and comprehensive VIKOR algorithm. Mater. Des. 47, 27–34.

Chang, C.L., 2010. A modified VIKOR method for multiple criteria analysis. Environ. Monit. Assess. 168, 339–344.

Chatterjee, P., Athawale, V.M., Chakraborty, S., 2009. Selection of materials using compromise ranking and outranking methods. Mater. Des. 30, 4043–4053.

Chatterjee, P., Athawale, V.M., Chakraborty, S., 2011. Materials selection using complex proportional assessment and evaluation of mixed data methods. Mater. Des. 32, 851–860.

Chen, S.J.J., Hwang, C.L., Beckmann, M.J., Krelle, W., 1992. Fuzzy Multiple Attribute Decision Making: Methods and Applications. Springer-Verlag, Secaucus, NJ.

Chiner, M., 1988. Planning of expert systems for materials selection. Mater. Des. 9, 195–203.

Chu, A.T.W., Kalaba, R.E., Spingarn, K., 1979. A comparison of two methods for determining the weights of belonging to fuzzy sets. J. Optim. Theory Appl. 27, 531–538.

Cicek, K., Celik, M., 2010. Multiple attribute decision making solution to material selection problem based on modified fuzzy axiomatic design-model selection interface algorithm. Mater. Des. 31, 2129–2133.

Dehghan-Manshadi, B., Mahmudi, H., Abedian, A., Mahmudi, R., 2007. A novel method for materials selection in mechanical design: combination of non-linear normalization and a modified digital logic method. Mater. Des. 28, 8–15.

Deng, H., Yeh, C.H., Willis, R.J., 2000. Inter-company comparison using modified TOPSIS with objective weights. Comput. Oper. Res. 27, 963–973.

Diakoulaki, D., Mavrotas, G., Papayannakis, L., 1995. Determining objective weights in multiple criteria problems: the critic method. Comput. Oper. Res. 22, 763–770.

Dieter, G.E., 1997. Overview of the Materials Selection Process. ASM Metals Handbook, Materials Selection and Design. ASM International, Materials Park, OH.

Doyle, J.R., Green, R.H., Bottomley, P.A., 1997. Judging relative importance: direct rating and point allocation are not equivalent. Organ. Behav. Hum. Decis. Process. 70, 65–72.

Durst, O., Ellermeier, J., Berger, C., 2008. Influence of plasma-nitriding and surface roughness on the wear and corrosion resistance of thin films (PVD/PECVD). Surf. Coatings Technol. 203, 848–854.

Edwards, W., 1977. How to use multiattribute utility measurement for social decision making. IEEE. Trans. Syst. Man. Cybern. 7, 326–340.

Edwards, K.L., 2005. Selecting materials for optimum use in engineering components. Mater. Des. 26, 469–473.

Farag, M.M., 1979. Materials and Process Selection in Engineering. Elsevier Science & Technology, London.

Farag, M.M., 1997. Materials Selection for Engineering Design. Prentice-Hall, New York.

Farag, M.M., 2002. Quantitative methods of materials selection. In: Kutz, M. (Ed.), Handbook of Materials Selection. John Wiley & Sons, London.

Favardin, P., Lepelley, D., Serais, J., 2002. Borda rule, Copeland method and strategic manipulation. Rev. Econ. Des. 7, 213–228.

Fayazbakhsh, K., Abedian, A., Manshadi, B.D., Khabbaz, R.S., 2009. Introducing a novel method for materials selection in mechanical design using Z-transformation in statistics for normalization of material properties. Mater. Des. 30, 4396–4404.

Figueira, J., Roy, B., 2002. Determining the weights of criteria in the ELECTRE type methods with a revised Simos' procedure. Eur. J. Oper. Res. 139, 317–326.

Figueira, J.R., Greco, S., Roy, B., 2009. ELECTRE methods with interaction between criteria: an extension of the concordance index. Eur. J. Oper. Res. 199, 478–495.

Fouladi, E., Fayazbakhsh, K., Abedian, A., 2010. Patch materials selection for ageing metallic aircraft structures using digital quantitative materials selection methods. 27th International Congress of the Aeronautical Sciences. Nice, France.

Giachetti, R.E., 1998. A decision support system for material and manufacturing process selection. J. Intell. Manuf. 9, 265–276.

Hajkowicz, S., Collins, K., 2007. A review of multiple criteria analysis for water resource planning and management. Water Resour. Manage. 21, 1553–1566.

Hambali, A., Sapuan, M.S., Ismail, N., Nukman, Y., 2010. Material selection of polymeric composite automotive bumper beam using analytical hierarchy process. J. Cent. South Univ. Technol. (English Edition). 17, 244–256.

Hambali, A., Sapuan, S.M., Rahim, A.S., Ismail, N., Nukman, Y., 2011. Concurrent decisions on design concept and material using analytical hierarchy process at the conceptual design stage. Concurrent Eng. Res. Appl. 19, 111–121.

Hobbs, B.F., Chankong, V., Hamadeh, W., Stakhiv, E.Z., 1992. Does choice of multicriteria method matter? An experiment in water resources planning. Water. Resour. Res. 28, 1767–1779.

Huang, J.J., Tzeng, G.H., Liu, H.H., 2009. A revised VIKOR model for multiple criteria decision making – the perspective of regret theory. Cutting-Edge Research Topics on Multiple Criteria Decision Making, 761–768.

Hwang, C.L., Lin, M.J., 1987. Group Decision Making Under Multiple Criteria: Methods and Applications. Springer-Verlag, Berlin.

Hwang, C.L., Yoon, K., 1981. Multiple Attribute Decision Making – Methods and Applications. Springer-Verlag, Berlin.

Jacobs, J.F., Van De Poel, I., Osseweijer, P., 2014. Clarifying the debate on selection methods for engineering: Arrow's impossibility theorem, design performances, and information basis. Res. Eng. Des. 25, 3–10.

Jahan, A., Edwards, K.L., 2013. VIKOR method for material selection problems with interval numbers and target-based criteria. Mater. Des. 47, 759–765.

Jahan, A., Edwards, K.L., 2015. A state-of-the-art survey on the influence of normalization techniques in ranking: improving the materials selection process in engineering design. Mater. Des. 65, 335–342.

Jahan, A., Ismail, M.Y., Sapuan, S.M., Mustapha, F., 2010. Material screening and choosing methods – A review. Mater. Des. 31, 696–705.

Jahan, A., Ismail, M.Y., Shuib, S., Norfazidah, D., Edwards, K.L., 2011a. An aggregation technique for optimal decision-making in materials selection. Mater. Des. 32, 4918–4924.

Jahan, A., Mustapha, F., Ismail, M.Y., Sapuan, S.M., Bahraminasab, M., 2011b. A comprehensive VIKOR method for material selection. Mater. Des. 32, 1215–1221.

Jahan, A., Bahraminasab, M., Edwards, K.L., 2012a. A target-based normalization technique for materials selection. Mater. Des. 35, 647–654.

Jahan, A., Mustapha, F., Sapuan, S.M., Ismail, M.Y., Bahraminasab, M., 2012b. A framework for weighting of criteria in ranking stage of material selection process. Int. J. Adv. Manuf. Technol. 58, 411–420.

Jee, D.H., Kang, K.J., 2000. A method for optimal material selection aided with decision making theory. Mater. Des. 21, 199–206.

Jeya Girubha, R., Vinodh, S., 2012. Application of fuzzy VIKOR and environmental impact analysis for material selection of an automotive component. Mater. Des. 37, 478–486.

Jiang, Y., Li, B., Tanabashi, Y., 2006. Estimating the relation between surface roughness and mechanical properties of rock joints. Int. J. Rock Mech. Min. Sci. 43, 837−846.

Karana, E., Hekkert, P., Kandachar, P., 2008. Material considerations in product design: a survey on crucial material aspects used by product designers. Mater. Des. 29, 1081−1089.

Karana, E., Hekkert, P., Kandachar, P., 2009. Meanings of materials through sensorial properties and manufacturing processes. Mater. Des. 30, 2778−2784.

Karande, P., Chakraborty, S., 2012. Application of multi-objective optimization on the basis of ratio analysis (MOORA) method for materials selection. Mater. Des. 37, 317−324.

Keeney, R.L., Raiffa, H., 1993. Decisions with Multiple Objectives: Preferences and Value Tradeoffs. Cambridge University Press, New York.

Lee, J.W., Kim, S.H., 2000. Using analytic network process and goal programming for interdependent information system project selection. Comput. Oper. Res. 27, 367−382.

Li, Y., Weston, N., Mavris, D., 2008. An approach for multi-criteria decision making method selection and development. 26th International Congress of the Aeronautical Sciences. Anchorage, AK.

Liao, T.W., 1996. A fuzzy multicriteria decision-making method for material selection. J. Manuf. Syst. 15, 1−12.

Lin, K.-P., Ho, H.-P., Hung, K.-C., Pai, P.-F., 2012. Combining fuzzy weight average with fuzzy inference system for material substitution selection in electric industry. Comput. Ind. Eng. 62, 1034−1045.

Ma, J., Fan, Z.P., Huang, L.H., 1999. A subjective and objective integrated approach to determine attribute weights. Eur. J. Oper. Res. 112, 397−404.

Maity, S.R., Chatterjee, P., Chakraborty, S., 2012. Cutting tool material selection using grey complex proportional assessment method. Mater. Des. 36, 372−378.

Majumdar, A., 2010. Selection of raw materials in textile spinning industry using fuzzy multi-criteria decision making approach. Fibers Polym. 11, 121−127.

Mardani, A., Jusoh, A., Zavadskas, E.K., 2015. Fuzzy multiple criteria decision-making techniques and applications − two decades review from 1994 to 2014. Expert Syst. Appl. 42, 4126−4148.

Michael, P., 2009a. Chapter 3 − Selecting materials. Materials Enabled Designs. Butterworth-Heinemann, Boston, MA.

Michael, P., 2009b. Chapter 10 − Detail design and testing. Materials Enabled Designs. Butterworth-Heinemann, Boston, MA.

Milani, A.S., Shanian, A., 2006. Gear material selection with uncertain and incomplete data. Material performance indices and decision aid model. Int. J. Mech. Mater. Des. 3, 209−222.

Milani, A.S., Shanian, A., Lynam, C., Saaty, T.L., 2012. An application of the analytic network process in multiple criteria material selection. Mater. Des.

Olson, D.L., Moshkovich, H.M., Schellenberger, R., Mechitov, A.I., 1995. Consistency and accuracy in decision aids: experiments with four multiattribute systems. Decis. Sci. 26, 723−747.

Opricovic, S., Tzeng, G.H., 2004. Compromise solution by MCDM methods: a comparative analysis of VIKOR and TOPSIS. Eur. J. Oper. Res. 156, 445−455.

Pomerol, J.C., Barba-Romero, S., 2000. Multicriterion Decision in Management: Principles and Practice. Springer, Amsterdam.

Pöyhönen, M., Hämäläinen, R.P., 2001. On the convergence of multiattribute weighting methods. Eur. J. Oper. Res. 129, 569−585.

Pratyyush, S., Jian-Bo, Y., 1998. Multiple Criteria Decision Support in Engineering Design. Springer, Berlin.

Ramalhete, P.S., Senos, A.M.R., Aguiar, C., 2010. Digital tools for material selection in product design. Mater. Des. 31, 2275–2287.

Ramík, J., Perzina, R., 2010. A method for solving fuzzy multicriteria decision problems with dependent criteria. Fuzzy Optim. Decis. Making. 9, 123–141.

Rao, R.V., 2006. A material selection model using graph theory and matrix approach. Mater. Sci. Eng. A. 431, 248–255.

Rao, R.V., 2008. A decision making methodology for material selection using an improved compromise ranking method. Mater. Des. 29, 1949–1954.

Rao, R.V., Davim, J.P., 2008. A decision-making framework model for material selection using a combined multiple attribute decision-making method. Int. J. Adv. Manuf. Technol. 35, 751–760.

Rao, R.V., Patel, B.K., 2010. A subjective and objective integrated multiple attribute decision making method for material selection. Mater. Des. 31, 4738–4747.

Rongxi, Z., Jianrong, X., Dayi, H., 2009. Approach of determining interval entropy weight based on the subjective preference of decision-maker and its application. Control and Decision Conference. Guilin.

Roth, R., Field, F., Clark, J., 1994. Materials selection and multi-attribute utility analysis. J. Comput. Aided Mater. Des. 1, 325–342.

Saaty, T.L., 1977. A scaling method for priorities in hierarchical structures. J. Math. Psychol. 15, 234–281.

Saaty, T.L., 1990. How to make a decision: the analytic hierarchy process. Eur. J. Oper. Res. 48, 9–26.

Sayadi, M.K., Heydari, M., Shahanaghi, K., 2009. Extension of VIKOR method for decision making problem with interval numbers. Appl. Math. Model. 33, 2257–2262.

Shanian, A., Savadogo, O., 2006a. ELECTRE I decision support model for material selection of bipolar plates for polymer electrolyte fuel cells applications. J. New Mater. Electrochem. Syst. 9, 191–199.

Shanian, A., Savadogo, O., 2006b. A material selection model based on the concept of multiple attribute decision making. Mater. Des. 27, 329–337.

Shanian, A., Savadogo, O., 2006c. A non-compensatory compromised solution for material selection of bipolar plates for polymer electrolyte membrane fuel cell (PEMFC) using ELECTRE IV. Electrochim. Acta. 51, 5307–5315.

Shanian, A., Savadogo, O., 2006d. TOPSIS multiple-criteria decision support analysis for material selection of metallic bipolar plates for polymer electrolyte fuel cell. J. Power Sources. 159, 1095–1104.

Shanian, A., Savadogo, O., 2009. A methodological concept for material selection of highly sensitive components based on multiple criteria decision analysis. Expert Syst. Appl. 36, 1362–1370.

Shanian, A., Milani, A.S., Carson, C., Abeyaratne, R.C., 2008. A new application of ELECTRE III and revised Simos' procedure for group material selection under weighting uncertainty. Knowl. Based Syst. 21, 709–720.

Shemshadi, A., Shirazi, H., Toreihi, M., Tarokh, M.J., 2011. A fuzzy VIKOR method for supplier selection based on entropy measure for objective weighting. Expert Syst. Appl. 38, 12160–12167.

Shirland, L.E., Jesse, R.R., Thompson, R.L., Iacovou, C.L., 2003. Determining attribute weights using mathematical programming. Omega. 31, 423–437.

Simanaviciene, R., Ustinovichius, L., 2010. Sensitivity analysis for multiple criteria decision making methods: TOPSIS and SAW. Procedia — Soc. Behav. Sci. 2, 7743—7744.

Swift, K.G., Raines, M., Booker, J.D., 2000. Case studies in probabilistic design. J. Eng. Des. 11, 299—316.

Thurston, D.L., Carnahan, J.V., 1992. Fuzzy ratings and utility analysis in preliminary design evaluation of multiple attributes. J. Mech. Des. — Trans. ASME. 114, 648—658.

Ullah, A.M.M., Harib, K.H., 2008. An intelligent method for selecting optimal materials and its application. Adv. Eng. Inform. 22, 473—483.

Van Kesteren, I.E.H., 2008. Product designers' information needs in materials selection. Mater. Des. 29, 133—145.

Vincke, P., 1992. Multicriteria Decision-Aid. John Wiley & Sons, New York.

Von Winterfeldt, D., Edwards, W., 1986. Decision Analysis and Behavioral Research. Cambridge University Press, Cambridge.

Voogd, H., 1983. Multicriteria Evaluation for Urban and Regional Planning. Pion Ltd., London.

Wallace, K., Burgess, S., 1995. Methods and tools for decision making in engineering design. Des. Stud. 16, 429—446.

Wang, Y.-M., Luo, Y., 2010. Integration of correlations with standard deviations for determining attribute weights in multiple attribute decision making. Math. Comput. Model. 51, 1—12.

Wang, Y.M., Parkan, C., 2006. A general multiple attribute decision-making approach for integrating subjective preferences and objective information. Fuzzy Sets Syst. 157, 1333—1345.

Ward, E., Hutton, B.F., 1994. SMARTS and SMARTER: improved simple methods for multiattribute utility measurement. Organ. Behav. Hum. Decis. Process. 60, 306—325.

Wu, H.H., 2002. A comparative study of using grey relational analysis in multiple attribute decision making problems. Qual. Eng. 15, 209—217.

Xu, X., 2004. A note on the subjective and objective integrated approach to determine attribute weights. Eur. J. Oper. Res. 156, 530—532.

Yeh, 2002. A problem-based selection of multi-attribute decision-making methods. Int. Trans. Oper. Res. 9, 169—181.

Yurdakul, M., Tansel, Ç.Y., 2009. Application of correlation test to criteria selection for multi criteria decision making (MCDM) models. Int. J. Adv. Manuf. Technol. 40, 403—412.

Zeleny, M., 1982. Multiple Criteria Decision Making. McGraw-Hill, New York.

Zhao, R., Neighbour, G., Deutz, P., Mcguire, M., 2012. Materials selection for cleaner production: an environmental evaluation approach. Mater. Des. 37, 429—434.

Zhou, P., Ang, B.W., Poh, K.L., 2006. Comparing aggregating methods for constructing the composite environmental index: an objective measure. Ecol. Econ. 59, 305—311.

Multiple objective decision-making for material and geometry design

6

Learning Aims

The overall aim of this chapter is to gain an overview of materials tailoring as well as material and design optimization. After carefully studying this chapter you should be able to understand:

- The difference between materials tailoring and materials selection.
- The significance of using the design of experiments (DoE) in the optimization of designed materials.
- The concept of response surface methodology for formulation and optimization.
- The importance of structural optimization combined with materials selection.

6.1   Material design/tailoring verses materials selection

The use of materials in different applications has evolved from "materials by chance" to "materials by design" as shown in Fig. 6.1 (Brechet and Embury, 2013). In the early days, the use of materials was based on its availability on site, such as using wood, stone, and some native metals. Later, to fulfill the needs and to improve the properties, optimization of particular classes of materials was carried out, first by experience and then by scientific methods, fundamentally with the use of metallurgy and afterwards materials science. Today, materials are mostly chosen by comparing and selecting from different previously optimized classes of materials. However, this might not be adequate for some engineering applications, which need specific materials to be carefully designed to meet complex multi-functional requirements.

Therefore, today, during the process of designing a product, materials are either selected or designed to meet the desired requirements. Materials selection usually deals with choosing the best material, from various material databases, that suit a specific product design requirement by considering multiple requirements and goals. Materials selection generally contains two important stages (Jahan et al., 2010): screening to discard inappropriate materials and ranking to obtain the best well-suited materials by considering several criteria based on the material properties, product geometry, and loading condition. For the screening of alternative

Multi-criteria Decision Analysis. DOI: http://dx.doi.org/10.1016/B978-0-08-100536-1.00006-0

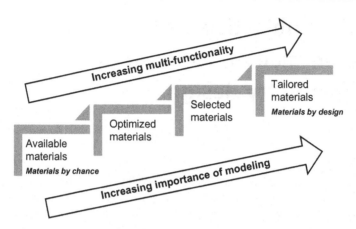

Figure 6.1 Four stages of materials evolution.

materials, different methods including Ashby's material charts approach (Ashby and Johnson, 2013) are available, and for ranking, multi-attribute decision-making (MADM) techniques are employed. However, for some applications, the achievement of design requirements might be restricted when selecting a material from existing materials. For instance, when a material is selected for a given application when all the suitable alternatives are metallic materials and therefore corrosion resistance is a key selection criterion. This is because there would always be an unavoidable amount of corrosion to consider when using metallic materials. Also, in the materials selection process, the final choice of material is often a compromise that has advantages and disadvantages, such as higher strength at the expense of higher weight.

To overcome the shortcomings in available materials, new materials can be designed, strategically to satisfy a particular application. Material design is tailoring material properties by compositional or structural variations in a controlled way and by configuring the processing route to achieve the desired requirements for a particular product design, and to improve the function in a way that is not obtainable with available materials. For example in development of a new ceramic-on-ceramic hip prosthesis, alumina and zirconia were first chosen. However, brittleness and low fracture toughness of alumina, and instability and phase transformation of zirconia restricted their use (Bal et al., 2006; Rahman et al., 2013). Now, the trend is towards designing hybrid materials; composites and functionally graded materials (FGMs) like zirconia toughened alumina, alumina matrix composite, alumina−silicon carbide−zirconia, hydroxyapatite−alumina−zirconia, and many others (Afzal et al., 2012; Askari et al., 2012; Bahraminasab et al., 2012). The conventional approach to the development of a new material is to change the processing route of an existing material to manipulate the material structure to change the properties and hence enhance the performance. The conventional approach therefore is basically a refined trial-and-error scheme. Nevertheless, the alternative new approach, which is more preferable, is that the desired performance is the driving element for determining the processing route to attain the material structure and the properties required for the desired performance. This would possibly be achieved by using advanced processing techniques and computer-based simulations and modeling. Indeed, modeling the multi-level structure of materials and the mechanical responses of materials under real loading conditions is a vital precursor. This enables the prediction of the behavior of materials and the identification of design variables of materials and their optimal values. Additionally, the implications of making changes in processing and structure can be readily evaluated and understood. Therefore, the relationship between structure, property, and the function has been identified and a basis for determining the manufacturing approach provided.

One thing that should be noted here is that the design of a new material will add design time, cost, and importantly risk to the new product development efforts. It is therefore to be used for the occasion either when there is no acceptable available material options or when the potential performance benefits are balanced by the increase in cost and time, particularly if the new material offers a product design with a competitive advantage in comparison with the products manufactured by other companies (Thompson, 2007; Jahan and Edwards, 2013). In other words, it should be borne in mind whether the product to be designed is cost-driven or performance-driven. The latter is important when the design process is carried out on products that vitally deal with human life (e.g., biomedical applications) when available materials cannot meet the requirements and then it is worth accepting the risks and longer time involved (see Table 6.1). For example, a material used for the femoral component of a total knee replacement requires some specific properties (in addition to good overall mechanical performance) to avoid the problem of aseptic loosening, which imposes revision surgery. Aseptic loosening is mainly related to the high elastic modulus of the implant relative to the bone (causing stress shielding and bone loss), excessive wear between the articular surfaces (causing osteolysis and bone loss), and poor bioactivity (causing weak anchorage with the bone and relative micro-motion). A material solution might be to provide multi-functions in the component through compositional and structural variations, that is, use of an FGM (Bahraminasab et al., 2013b). Fig. 6.2 shows a femoral component, the leading causes of aseptic loosening and the FGM material solution. The suggested FGM is functionally graded from titanium (Ti) with a large amount of porosity at the upper surface where the component interfaces the bone, to alumina ceramic (Al_2O_3) with zero porosity in the articulating surface. The reason for choosing titanium is the bioactivity and moderate elastic modulus. The porosity further reduces the modulus and allows penetration of bone cells into the implant, which provides a strong connection with the bone. The alumina ceramic is a high wear resistant material and is therefore suitable for the articular surfaces. The effectiveness of the proposed new material has been evaluated by computer simulation and optimization processes (Bahraminasab et al., 2013b; Bahraminasab et al., 2014a). As described by Ashby (2005), in addition to the design of the material, product design and the manufacturing approach are also required to be adjusted to fulfill

Table 6.1 Comparison of materials selection and material design

	Materials selection	Material design
Cost	Low	High
Meeting design requirements	Sometimes involves compromises and sometimes involves restrictions in satisfying requirements due to deficiencies of existing materials	Can satisfy all requirements, even the conflicting ones
Design process	Simple (well-established procedures and fewer design decisions regarding the choice of material)	Complex (requires sophisticated modeling and testing)
Timeline	Shorter (all available materials have already been evaluated and databases exist)	Longer (usually contains simulation for an earlier design and optimization, and then confirmation experiments for mechanical evaluations)
Tools and techniques	Usually Quality Function Deployment (QFD) for translating design requirements, Ashby's approach for screening, and MADM analysis techniques for ranking	Needs different design tools, including: Finite Element Analysis (FEA), DoE, QFD, Multi-objective Decision-Making (MODM) analysis, and mechanical experiments/tests for validation purposes

the desired requirements entirely. Fig. 6.3, shows an example of such a relationship for the design of the aforementioned FGM. Another example of tailored materials is composites, which require the design of component and manufacturing process to be considered together early in the design process (Sapuan and Mansor, 2014). This was implemented for the development of an automotive vehicle bumper beam made of polymer matrix composites (Hambali, 2009). The component design was conducted using computer-aided design (CAD) system, and materials selection, concept design selection, and manufacturing process selection were carried out using the Analytical Hierarchy Process (AHP). Table 6.1 demonstrates the similarities and differences between the materials selection and material design processes.

Fig. 6.4 shows materials screening, choosing, and design steps. This starts with identifying the functional requirements of a product and engineering analysis of materials used for similar applications. Based on the desired functions, improper materials are screened out by applying Ashby's approach. Depending on

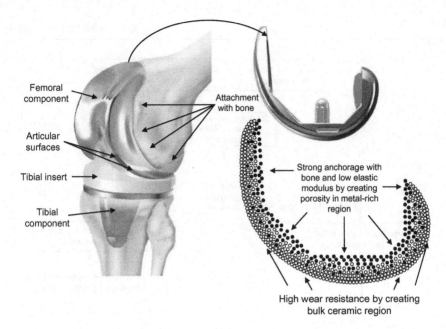

Figure 6.2 Femoral component, leading causes of aseptic loosening and the material solution for a total knee replacement.

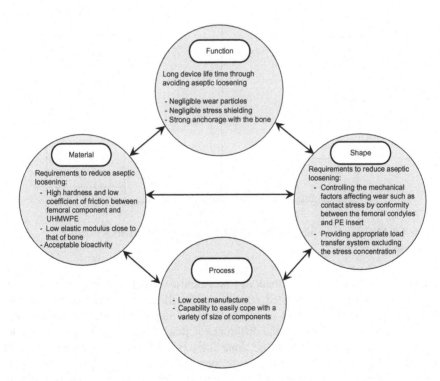

Figure 6.3 An example of relationship between material, design, process, and performance: The femoral component of a total knee replacement.

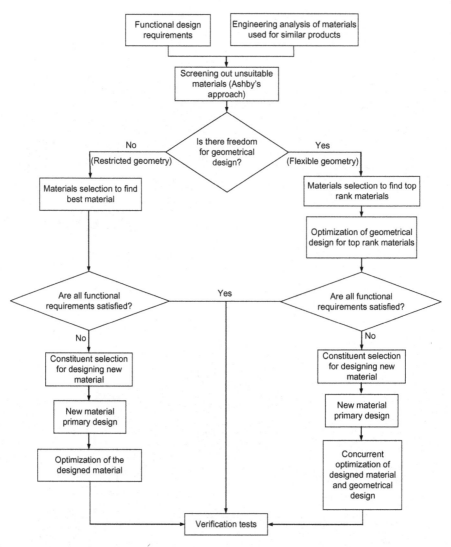

Figure 6.4 Materials screening, choosing, and design steps: Materials selection versus material design.

whether there is geometrical freedom (the extent to which a product's component shapes can possibly be adjusted whilst still fulfilling the design requirements), two strategies can be applied in which the first step is material selection using MADM techniques. However, the difference is that when the geometry is flexible, optimization of dimensions can be done for top ranked materials using MODM and FEA techniques to provide more performance improvement. When this process cannot yield all the design functional requirements, material design should be established. This approach contains constituents' selection, earlier

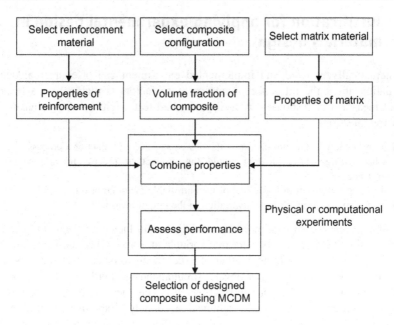

Figure 6.5 A framework for selecting of designed composite material.

design of the material, and an optimization of the material to obtain the best function and where the geometry is flexible it is better to be done along with geometry optimization concurrently.

It can be seen from Fig. 6.4 that material design inherently involves materials selection: by applying MADM techniques either for constituent materials selection or prior to design of material to ensure that currently available materials are not able to meet the functional requirements. An example of this may appear in the design of a composite material as shown in Fig. 6.5. In the case of composites, candidate reinforcement and matrix materials, and composite configuration (fibers, particles, etc.) have to be initially selected, and then their relative properties combined and performance assessed to select a new composite material. Therefore, to maximize determining the most suitable material, it is necessary to use a database of material properties that consists of the properties of both existing materials and newly developed or designed materials. The latter materials may not necessarily be proven, but theoretically calculated properties of promising materials derived from simulation-based materials design. Therefore, it is possible to interrogate judiciously a comprehensive database of material properties to search for a material with the most favorable properties (Curtarolo et al., 2013). Therefore, by having thorough knowledge of materials properties, a proper risk assessment can then be made when the selection needs to be conducted between well-established materials and novel materials. Obviously, using MCDM methods can increase the potential of detecting all appropriate materials, leading to the longer lasting success of a technological sector.

6.2   Justification for applying experimental design in materials design

A systematically planned and implemented experiment can offer a great deal of information about the influence of one or more factors (variables) on a response (output measure) or responses. A well performed test might be able to answer the following questions:

- What are the key influential factors in the design stage of a product or a process?
- At what settings (factor levels) would the product or process provide satisfactory performance?
- What setting will result in less effect of uncontrollable factors (noise factors)?
- What settings would result in less variation in the output measure?

Most experiments are conducted by holding certain factors constant and changing the levels of one factor, that is, alter one variable at a time. This one-factor-at-a-time (OFAT) approach needs a large number of experiments to be conducted and becomes ineffective when varying factor levels simultaneously. Furthermore, it does not designate the relationships and interactions between the variables. Therefore, the optimal conditions might not be achieved accurately. Experimental design, also known as DoE, is an efficient statistical method of dealing with the planning, conducting, analyzing, and interpreting of controlled tests to assess the factors which affect the value of a response or responses. DoE is employed for designing an experiment that requires the minimum number of tests to be carried out and still obtain the necessary information. It can consider a large number of factors (f) at a fixed number of levels (n) and indicate the factor levels that will concurrently meet a set of desired goals. DoE is usually used with three aims (Robinson et al., 2004; Ilzarbe et al., 2008; Myers et al., 2009):

- Screening to understand the effect of factors on the response variable and narrow the field of under assessment variables.
- Response Surface Methodology (RSM) to provide both an understanding of response behavior and to find the optimum point by modeling the response (explained briefly in the following section).
- Robust Parameter Design (RPD) to find the settings of the controllable design factors that minimize the variability in response caused by noise factors.

DoE includes approaches like Taguchi's orthogonal matrices, full factorial designs, fractional factorial designs, central composite designs, Placket-Burmann designs, optimal designs, Box-Behnken designs, etc. The most conservative design in DoE approaches is full factorial designs, because all possible combinations of the factor settings (including both continuous and categorical factors) are examined, thus there is little ambiguity. In full factorial design, the number of required experiments is n^f which grows exponentially in the number of factors. Hence full factorial designs become impractical once there are large number of factors. Nevertheless, by applying fractional factorial design, the number of experiments required can be considerably reduced whilst still extracting the essential information. DoE as a quick and cost-effective technique has been used in various fields including mechanical engineering (Little, 2002), chemical engineering (Lazic, 2004), and manufacturing (Ilzarbe et al., 2008).

In addition to real (physical) experimentation, DoE recently has been used in conjunction with computational (numerical) experiments. In the former, real data are obtained by conducting experimentation in the laboratory or manufacturing plant, and in the latter, simulated data are generated using computer software. Fig. 6.6 compares the use of DoE in engineering applications by real and simulated experiments (Ilzarbe et al., 2008). One area in which computer simulations are most frequently utilized is biomedical engineering, where the development of new implant materials is associated with high cost, complicated process of implant manufacturing, and time consuming in vitro and in vivo testing. Computer simulations are getting progressively popular for running experiments associated with products and process development. The important advantages of using simulations is the ability to study a lot of factor settings, and in reducing the time and cost of physical testing. Also, if the experiments are destructive, the benefit of using computer modeling becomes more obvious. The other noteworthy issue is that when applying computer-based simulations, there would be no actual experimental error and consequently replication of the design points is not necessary.

Therefore, DoE can be usefully employed in the design stage of new hybrid materials either with real experiments or along with FEA as a simulation tool, in which there exists several variables to be evaluated and to be optimized with respect to the performance of the material (responses). Therefore, the first step is to determine the variables of the hybrid material such as volume fraction of constituents, configuration, orientation, distribution, and amount of porosity (Table 6.2), and the responses which are the performance expected from the new material in a specific application. The next step is to identify the ranges of variation for the factors at which the experiments are going to be run in order to find the material response.

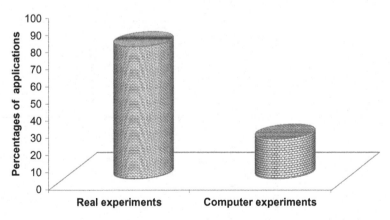

Figure 6.6 Comparing of using DoE in engineering applications by real and simulated experiments.

Table 6.2 Examples of using DoE for design of new materials

Reference	Designed material	Type of experimental design	Material design variables	Responses	Types of experiment	Objective
Pelegri and Tekkam (2003)	Cross ply carbon-epoxy laminated composite	Taguchi approach	Width, length and thickness Specimen delamination length and stacking sequence	Mode I delamination fracture toughness	Real	Optimization
Milani et al. (2008)	Graphite-epoxy laminated composite	OFAT, RSM, Bayesian DoE-based methodology	Ply angles	Tsai-Hill failure index	Simulation	Optimization
Cihan et al. (2012)	Concrete	Fractional factorial design and D-optimal design	Water/cement ratio, Cement content, Compressive strength of cement, Fineness modulus of the aggregate mix, Fineness content of the aggregate mix, Admixture dosage, Aggregate type	Compressive strength of concrete	Real	Optimization
Bahraminasab et al. (2014a)	Porous metal-ceramic FGM	Central composite design	One control parameter for volume fraction of composition and two control parameters for porosity	Five responses related to the function of material for reduction of aseptic loosening, the main cause of reduction of implant failure	Simulation	Optimization

Then, at the end, statistical analysis is carried out to indicate the influential factors and the values of variables at which the material performs better. This strategy has been used in several studies, four of which are summarized in Table 6.2.

Despite the significant progress being made in applying DoE, it has been underutilized in some areas, especially in computational experiments such as in biomechanics (Taylor and Prendergast, 2015). Furthermore, there is little use of advanced strategies in which statistical concepts are deeply implemented for experimentation. To apply this technique, comprehensively, it is necessary that statistical concepts and technical knowledge in the fields such as material science and mechanical engineering are synthesized.

6.3   RSM in multi-objective mathematical modeling of material design

RSM, as a subcategory of DoE methods, is an efficient statistical tool for design, development, improvement, and optimization of product designs and processes (Myers et al., 2009). The method has been extensively applied in industry where several input variables (factors) potentially affect some output measures or quality characteristics (responses) of a product or process. RSM can identify the optimal combination of input variables that will result in achievement of the desired response for a certain measurement and describe the response at the vicinity of the optimum point. Most design problems, including new material design, involves more than one response to be simultaneously considered (Deka et al., 2005; Pelletier and Vel, 2006; Goupee and Vel, 2007; Lin et al., 2009; Sadollah and Bahreininejad, 2011; Bahraminasab et al., 2014a).

When considering the actual application and operating conditions, most material design problems are multi-criteria or multi-objective in nature. This can be found in the optimization of hybrid materials such as FGMs for orthopedic prostheses (Bahraminasab et al., 2014a), dental implants (Lin et al., 2009; Sadollah and Bahreininejad, 2011), and high-temperature applications (Goupee and Vel, 2007). In addition to hybrid materials, optimization of conventional materials like engineering alloys, also deals with several objectives. As an example, the optimizing of steel alloys for crankshafts; yield strength and elongation are the objectives of the design problem. The multi-objective optimization can be tackled by many different techniques in an engineering design context, which have been summarized in a study conducted by Andersson (2000). The problems with multi-objectives can be readily dealt with by the use of RSM due to its capability in the formulation of responses and multi-objective optimization. RSM usually constructs graphical displays (either in three-dimensional space or as contour plots) of the responses with the input variables, which aid in visualizing the response behavior with each level of a factor and with a combination of factors. This makes the optimization process very straightforward. A graphical display may provide a response surface lying above the two of variables. Therefore, by inspection of the plot, the values of

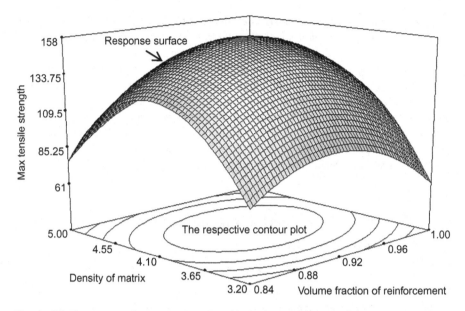

Figure 6.7 Response surface graph for a hypothetical composite material.

the variables at which the response is maximized or minimized are obtained. This graphical perspective of the problem implies the term RSM. Fig. 6.7 shows an example of response surface graph for a hypothetical composite material.

In general, the response (y) depends upon the controllable input variables ($\xi_1, \xi_2, \ldots, \xi_k$) where the relationship can be expressed by:

$$y = f(\xi_1, \xi_2, \ldots, \xi_k) + \varepsilon \tag{6.1}$$

The form of the true response function f (the relationship between the response and the input variables) is not known and possibly very complex, and ε is the random statistical error caused by other sources of variability including the measurement errors, other causes of variation which are intrinsic in the system, the influence of other possibly unknown variables, etc. The ε has a normal distribution with a mean of zero and constant variance, which is usual in statistical modeling. The input variables $\xi_1, \xi_2, \ldots, \xi_k$ in Eq. (6.1) are typically expressed in the natural measurement units and termed as natural variables. However, in RSM, it is common to convert the natural variables to the coded variables ($X_1, X_2, \ldots, X_k$). The coded variables are dimensionless which have a mean zero and the same spread or standard deviation. Thus, the Eq. (6.1) can be transformed to:

$$\eta = f(X_1, X_2, \ldots, X_k) + \varepsilon \tag{6.2}$$

In general, these mathematical models are polynomials with an unidentified structure (coefficients). First-order (linear) and second-order (quadratic) models are applied in most problems. Eqs. (6.3) and (6.4) show a coded first-order model and a coded full quadratic model, respectively.

$$\eta = \beta_0 + \beta_1 X_1 + \beta_2 X_2 + \beta_{12} X_1 X_2 + \varepsilon \tag{6.3}$$

$$\eta = \beta_0 + \beta_1 X_1 + \beta_2 X_2 + \beta_3 X_3 + \beta_{11} X_1^2 + \beta_{22} X_2^2 + \beta_{33} X_3^2 + \beta_{12} X_1 X_2$$
$$+ \beta_{13} X_1 X_3 + \beta_{23} X_2 X_3 + \varepsilon \tag{6.4}$$

The unknown regression parameters β_1, β_2, β_3 are coefficients for main effects, β_{11}, β_{22}, β_{33} are coefficients for quadratic main effects, and β_{12}, β_{13}, β_{23} are coefficients for two factor interaction effects, which can be obtained by the least squares method. In Eq. (6.5), the RSM can be expressed in terms of the parameter estimates as:

$$\hat{\eta} = b_0 + b_1 X_1 + b_2 X_2 + b_3 X_3 + b_{11} X_1^2 + b_{22} X_2^2 + b_{33} X_3^2 + b_{12} X_1 X_2$$
$$+ b_{13} X_1 X_3 + b_{23} X_2 X_3 + \varepsilon \tag{6.5}$$

where the caret (^) designates that the predicted response is an approximation according to the fitted RSM. A second-order model is normally applied to estimate the response when the first-order model is no longer acceptable. In order to obtain proper appreciation about the design process of a product as the investigation proceeds, sequential steps are required to take in RSM. Fig. 6.8 shows the general required sequential steps in the optimization process using RSM.

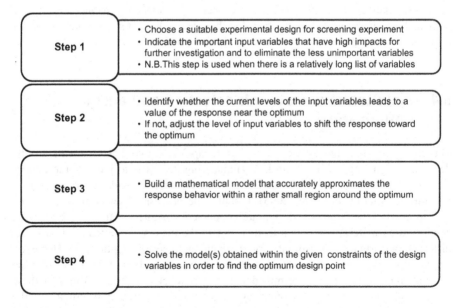

Figure 6.8 General required sequential steps in optimization process using RSM.

The design of a new hybrid material may provide improved properties, but nonetheless does not necessarily guarantee the best performance. This means that the performance of hybrid materials does not merely depend upon the relative amount of material constituents and their individual properties alone, but is directly related to using the constituents in the most optimal manner. To achieve this goal, therefore, there is a need for developing the optimum design of the hybrid material. Today, because of simulation-based design of new hybrid materials, optimization of these materials, particularly using RSM, is coupled with computer simulation tools without the need to build a physical prototype. As a consequence, this accelerates the development of the material and hence reduces the cost of trial-and-error experiments. Simulation-based optimization techniques coupled with RSM can ensure the best composition and structure of the designed material. This will also impart the best functions in the product and help to recognize the most sensitive design variables in order to establish an improved control plan throughout the manufacturing stage. As already mentioned, the design of hybrid materials usually involves more than one design objective that often conflict and must be simultaneously satisfied. The trade-offs between design objectives and exploration of design options requires the optimization problem to be formulated with multiple objectives. Fig. 6.9 proposes the required steps for a new hybrid material design using computational experiments and MODM techniques. It should be noted that for most multi-objective optimization problems, it is impossible to reach a single solution that concurrently optimizes each objective. Searching for solutions results in achieving some points so that, when attempt to enhance one objective, other objectives may consequently suffer.

The regression model (RSM) enables material engineers to predict the properties of material/performance indices with respect to influencing parameters. These problems are characterized by the fact that they do not have a single solution, and it is necessary to find the Pareto front, as explained previously in Chapter 4.

6.4   Simultaneous materials selection and design optimization

Materials selection and structural design (shape) are two key elements of the detailed design stage. The key task in engineering design is to identify the dimensions of the component and the materials from which it is made in order for it to perform acceptably (Edwards, 2005). Usually, the process of selecting the material and optimizing the structural design are conducted separately. Most of the research conducted on materials selection tends to select the most suitable material based on different performance constraints for a given engineering application. However, structural optimization of mechanical components is usually carried out to find the optimal dimensions for a determined material (Qiu et al., 2013). Although these two aspects are important in their own right, their integration is crucial in the successful design of new engineering products (Edwards, 2011) because the design

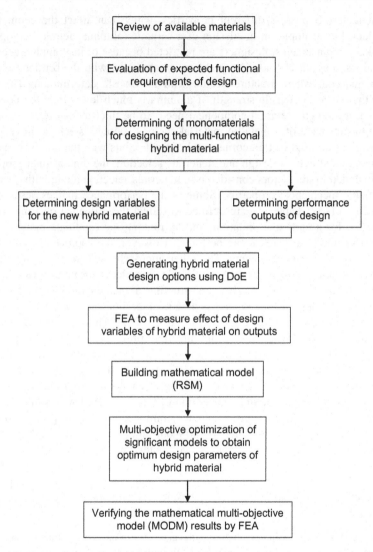

Figure 6.9 Proposed steps for new hybrid material design using computational experiments and MODM techniques.

geometry of a mechanical component depends on the materials selected for its manufacture, and vice versa. Therefore, without considering the geometry of the product, the optimum design may not be attained. However, the specific structure and characteristics of the material might not be completely utilized, leading to wastage. Alternatively, some of the material properties may not fulfill particular structural requirements and this may impose the specific dimension to be enlarged, leading to a thick or heavy weight component. Meanwhile, when optimizing the geometry for a determined material, the solution spaces are confined for a given material as the different materials possess different properties, which can affect the geometry of the product. For example, in the design of a ceramic (alumina) femoral component in total knee replacements, designers are restricted because of its complex geometry (Bahraminasab et al., 2013a). The problem is associated with the brittleness of the ceramic material, where fracture occurs with no ductile deformation. The other issue is that since the tensile strength of alumina is much lower than the compressive or shear strength, femoral component needs to be thicker than the corresponding component made of a metallic material such as stainless steel, to prevent high tensile stresses at the resection corners and contact points with the polymer insert.

To integrate structural design and materials selection, the dimensional properties can be included in the factors considered in materials selection along with chemical, physical, and mechanical properties. Some of the performance indices described by Ashby (2000) (see Chapter 3), are defined to reflect the influence of a material on the structure for a particular situation, and by ranking the candidate materials with these indices, the best material can be found. However, the approach is useful only for rather simple design problems. The other approach used for the integration of these two elements, is to first use Ashby's selection charts for preliminary identification of candidate materials. Then constructing a mathematical formulation for structural optimization based on the geometrical design variables; the objectives, constraints, and particular loading condition using FEA. This can then be linked to an optimization system including optimization programs (optimizer), problem-dependent analyzer and interface programs to find the optimum structural solution for each material (Ermolaeva et al., 2002). However, the method is complicated and requires advanced computer programming to analyze the optimal structure for each candidate material. The implementation of DoE and RSM with simulated data, as explained in previous sections, facilitates combined structural optimization with materials selection, which is easier to carry out and with a small number of computer runs. Meanwhile, statistical analysis computer software such as *Design-Expert* program, readily enables both formulating and solving the design objective(s) for each candidate material. In fact, the following steps can be used to accomplish an exact structural optimization and materials selection simultaneously (Bahraminasab et al., 2014b):

- Applying DoE and FEA in conjunction with each other to analyze the interactions, and to determine the direction of enhancement in goals with respect to design variables.
- Finding the prime optimized geometry of each candidate material through single or multi-objective optimization using a statistical software package with the ability of solving the formulated objective(s), for example, Design-Expert computer program.
- Making a database of optimal designs and selecting the final design using MADM techniques.

It should be noted that in the optimization process, changing some of the dimensions is not feasible in order to meet the necessary functional requirements or assembly constraints such as those related to the interfacing geometries of two contacting surfaces, which slide and roll against each other. Although, the other changeable dimensions are taken as the geometrical design variables, which are

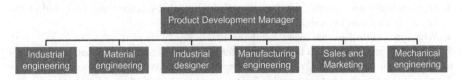

Figure 6.10 Required fields for successful product development.

continuous variables with the respective continual ranges indicated by considering the design constraints. The candidate materials are taken as discrete/categorical variables, which contain all the possible materials, including those screened from the material database, by Ashby's charts for example, or those promising materials previously tailored and optimized. In concurrent structural optimization and materials selection, the feasible geometries are optimized with respect to the objective(s) for each material. Some solutions will be obtained, which should be tested by either real experiments or by simulation to verify the optimum points.

The integration of materials selection and design optimization provides improved performance of a product, although further improvement can be made when the manufacturing constraints are also concurrently taken into account. This is of especially important for the development of hybrid materials due to the establishment of new or modified processing route. Therefore, for contemporary product development, the involvement of a team approach is necessary and this requires skills from different fields of study coordinated by a product development manager. Fig. 6.10 shows the required fields for successful product development.

6.5 Summary and conclusions

Engineering designers have to consider many different objectives, including material design, when necessary for new product development. For such design problems, different MODM methods, as well as MADM methods, are available in the literature. Therefore, the application of multi-objective optimization in materials engineering and design is emphasized, rather than explaining MODM methods. Also, RSM, a subcategory of DoE methods, can be well-fitted to the mathematical modeling of materials' performance, and is therefore considered to be an important integral part of the process.

Review questions

1. Explain why a designer might prefer to design a new material rather than select an available material.
2. Compare and contrast the steps involved in material design and product design.

3. Discuss, using an example, the four stages of materials evolution, that is, materials by chance to materials by design.
4. What types of decision are made during material design?
5. Name the variables that can be considered for the design of a composite material.
6. How do you think that DoE methods can improve the quality of design process?
7. What do you think about the possible limitations of mathematical modeling for optimizing material and design?
8. Using an example, try to explain the weak points of a design when using separate materials selection and geometry optimization processes.
9. What skills and knowledge are necessary for a design team to be successful?

References

Afzal, M.A.F., Kesarwani, P., Reddy, K.M., Kalmodia, S., Basu, B., Balani, K., 2012. Functionally graded hydroxyapatite-alumina-zirconia biocomposite: Synergy of toughness and biocompatibility. Mater. Sci. Eng. C. 32, 1164−1173.

Andersson, J., 2000. A survey of multiobjective optimization in engineering design. Department of Mechanical Engineering. Linkoping University, Linkoping, Sweden, Technical Report No: LiTH-IKP.

Ashby, M., 2000. Multi-objective optimization in material design and selection. Acta Mater. 48, 359−369.

Ashby, M.F., 2005. Materials Selection in Mechanical Design. Butterworth-Heinemann, Amsterdam.

Ashby, M.F., Johnson, K., 2013. Materials and Design: The Art and Science of Material Selection in Product Design. Butterworth-Heinemann, Amsterdam.

Askari, E., Mehrali, M., Metselaar, I.H.S.C., Kadri, N.A., Rahman, M., 2012. Fabrication and mechanical properties of $Al_2O_3/SiC/ZrO_2$ functionally graded material by electrophoretic deposition. J. Mech. Behav. Biomed. Mater. 12, 144−150.

Bahraminasab, M., Sahari, B.B., Edwards, K.L., Farahmand, F., Arumugam, M., Hong, T.S., 2012. Aseptic loosening of femoral components − a review of current and future trends in materials used. Mater. Des. 42, 459−470.

Bahraminasab, M., Sahari, B.B., Edwards, K.L., Farahmand, F., Arumugam, M., 2013a. Aseptic loosening of femoral components-materials engineering and design considerations. Mater. Des. 44, 155−163.

Bahraminasab, M., Sahari, B.B., Edwards, K.L., Farahmand, F., Hong, T.S., Naghibi, H., 2013b. Material tailoring of the femoral component in a total knee replacement to reduce the problem of aseptic loosening. Mater. Des. 52, 441−451.

Bahraminasab, M., Sahari, B., Edwards, K., Farahmand, F., Hong, T.S., Arumugam, M., et al., 2014a. Multi-objective design optimization of functionally graded material for the femoral component of a total knee replacement. Mater. Des. 53, 159−173.

Bahraminasab, M., Sahari, B., Edwards, K., Farahmand, F., Jahan, A., Hong, T.S., et al., 2014b. On the influence of shape and material used for the femoral component pegs in knee prostheses for reducing the problem of aseptic loosening. Mater. Des. 55, 416−428.

Bal, B.S., Garino, J., Ries, M., Rahaman, M.N., 2006. Ceramic materials in total joint arthroplasty. Semin. Arthroplasty. 17, 94−101.

Brechet, Y., Embury, J.D., 2013. Architectured materials: Expanding materials space. Scr. Mater. 68, 1–3.

Cihan, M.T., Guner, A., Yuzer, N., 2012. Response surfaces for compressive strength of concrete. Constr. Build. Mater. 40, 763–774.

Curtarolo, S., Hart, G.L., Nardelli, M.B., Mingo, N., Sanvito, S., Levy, O., 2013. The high-throughput highway to computational materials design. Nat. Mater. 12, 191–201.

Deka, D.J., Sandeep, G., Chakraborty, D., Dutta, A., 2005. Multiobjective optimization of laminated composites using finite element method and genetic algorithm. J. Reinf. Plast. Comp. 24, 273–285.

Edwards, K., 2005. Selecting materials for optimum use in engineering components. Mater. Des. 26, 469–473.

Edwards, K.L., 2011. Materials influence on design: A decade of development. Mater. Des. 32, 1073–1080.

Ermolaeva, N.S., Kaveline, K.G., Spoormaker, J.L., 2002. Materials selection combined with optimal structural design: Concept and some results. Mater. Des. 23, 459–470.

Goupee, A.J., Vel, S.S., 2007. Multi-objective optimization of functionally graded materials with temperature-dependent material properties. Mater. Des. 28, 1861–1879.

Hambali, A., 2009. Selection of Conceptual Design Using Analytical Hierarchy Process for Automotive Bumper Beam under Concurrent Engineering Environment. Universiti Putra Malaysia, Serdang.

Ilzarbe, L., Alvarez, M.J., Viles, E., Tanco, M., 2008. Practical applications of design of experiments in the field of engineering: A bibliographical review. Qual. Reliab. Eng. Int. 24, 417–428.

Jahan, A., Edwards, K.L., 2013. Weighting of dependent and target-based criteria for optimal decision-making in materials selection process: Biomedical applications. Mater. Des. 49, 1000–1008.

Jahan, A., Ismail, M., Sapuan, S., Mustapha, F., 2010. Material screening and choosing methods – A review. Mater. Des. 31, 696–705.

Lazic, Z.R., 2004. Design of Experiments in Chemical Engineering: A Practical Guide. Wiley-VCH, Weinheim.

Lin, D., Li, Q., Li, W., Zhou, S., Swain, M.V., 2009. Design optimization of functionally graded dental implant for bone remodeling. Composites Part B: Eng. 40, 668–675.

Little, R., 2003. Mechanical Reliability Improvement: Probability and Statistics for Experimental Testing. Marcel Dekker, New York.

Milani, A., Wang, H., Frey, D., Abeyaratne, R., 2008. Evaluating three DOE methodologies: Optimization of a composite laminate under fabrication error. Qual. Eng. 21, 96–110.

Myers, R.H., Montgomery, D.C., Anderson-Cook, C.M., 2009. Response Surface Methodology: Process and Product Optimization Using Designed Experiments. John Wiley & Sons, New York.

Pelegri, A.A., Tekkam, A., 2003. Optimization of laminates fracture toughness using design of experiments and response surface. J. Comp. Mater. 37, 579–596.

Pelletier, J.L., Vel, S.S., 2006. Multi-objective optimization of fiber reinforced composite laminates for strength, stiffness and minimal mass. Comput. Struct. 84, 2065–2080.

Qiu, L.-M., Sun, L.-F., Liu, X.-J., Zhang, S.-Y., 2013. Material selection combined with optimal structural design for mechanical parts. J. Zhejiang Univ. Sci. A. 14, 383–392.

Rahman, H.S.A., Choudhury, D., Osman, N.A.A., Shasmin, H.N., Abas, W.A.B.W., 2013. In vivo and in vitro outcomes of alumina, zirconia and their composited ceramic-on-ceramic hip joints. J. Ceram. Soc. Jpn. 121, 382–387.

Robinson, T.J., Borror, C.M., Myers, R.H., 2004. Robust parameter design: a review. Qual. Reliab. Eng. Int. 20, 81–101.

Sadollah, A., Bahreininejad, A., 2011. Optimum gradient material for a functionally graded dental implant using metaheuristic algorithms. J. Mech. Behav. Biomed. Mater. 4, 1384–1395.

Sapuan, S., Mansor, M.R., 2014. Concurrent engineering approach in the development of composite products: A review. Mater. Des. 58, 161–167.

Taylor, M., Prendergast, P.J., 2015. Four decades of finite element analysis of orthopaedic devices: Where are we now and what are the opportunities? J. Biomech. 48, 767–778.

Thompson, S.C., 2007. Material Design vs Material Selection: A Trade-Off between Design Freedom and Design Simplicity. Georgia Institute of Technology, Atlanta, GA.

Case studies of materials selection and design

7

Learning Aims

The overall aim of this chapter is to demonstrate the application of the methods described in the previous chapters for materials design, and materials and design selection. After carefully studying this chapter you should be able to understand:

- how to rank materials with cost, benefit, and target design criteria
- the concept of multi-objective material design optimization
- the materials selection and geometrical optimization process
- the advantages of quality function deployment (QFD) in the materials selection process
- how to rank materials and design scenarios with data uncertainty in design attributes
- the importance of materials selection strategy for multiple components
- the application of Pareto front diagram in materials selection
- the use of group decision-making by a team of engineers and designers evaluating qualitative criteria in the materials selection process.

7.1 Rationale for the case studies

It has already been explained that there are a very large number of materials, and associated materials processes, available to the designer. To select suitable materials for a particular application necessitates the simultaneous consideration of many conflicting criteria. Apart from the simplest of products this is normally a difficult problem-solving activity. The choice of materials is often restrained (by keeping to what you know) and assumptions made (because of restricted information) and approximations used (since the analysis is difficult) to achieve workable solutions. This stifles the uptake of different materials (to a particular designer) or new materials (for all designers), and when used are often based on experience only leading to inappropriate deployment, for example, anisotropic materials treated as if they are isotropic materials, leading to possible underutilization of the materials' performance or premature failure of a component.

The methods described in the preceding chapters have been developed to assist designers cope with complex (real world) materials selection and design problems. However, to help understand and appreciate the capability of the methods, worked examples and case studies are invaluable. Recent publications demonstrate the issues pertaining to the optimal selection of materials by incorporating challenging

Multi-criteria Decision Analysis. DOI: http://dx.doi.org/10.1016/B978-0-08-100536-1.00007-2

case studies such as the design of bipolar plates for a polymer electrolyte membrane fuel cell (Shanian and Savadogo, 2006), flywheel (Jee and Kang, 2000), cryogenic storage tank (Farag, 1997), spar for the wing of aircraft (Dehghan-Manshadi et al., 2007), and loaded thermal conductor (Shanian et al., 2008). These and other case studies have significant impact on the future development of improved decision support systems for materials selection. The detailed examples or case studies that follow have been carefully chosen to demonstrate the scope and explain the materials design and selection methods described in the preceding chapters.

7.2 Materials selection and design for biomedical implants

Biomaterials are artificial or natural materials that replace diseased or damaged organic systems to restore form and function and to provide a pain-free life for patients. Biomaterials are used in different parts of the human body such as artificial valves in the heart, stents in blood vessels, replacement implants in shoulders, knees, hips, elbows, ears, and orodental structures (Ramakrishna et al., 2001; Park and Bronzino, 2003). In the case of biomedical implants, the increasingly ageing population is continually raising the expectations and demands for creating new and improved products, economically and in high volumes. As a consequence, developing new and improved biomedical implants is seen as a complex design problem-solving activity and, in conjunction with demanding manufacturing constraints, utilizing the most appropriate materials (and materials combinations) presents many unique challenges. The selection and design of a new material for a biomedical implant application requires knowledge both on the functional requirements of the implant, like any other engineering product, and on the interaction of the material with the human body. In fact, the selected/designed implant material should provide the physiologic role of the replaced organ, and should last for a long time to fulfill that role. Among the many biomedical implants, total hip and total knee prostheses are two commonly replaced orthopedic implants (Kurtz et al., 2007), which despite the advances made, still lack sufficient innovative material solutions for long-term use in the human body.

The material properties are site-specific in the body and the requirements may vary from one application to another, however, total hip and knee prostheses require almost the same considerations. The hip is basically a ball-and-socket joint and the knee is a hinge type joint having six degrees of freedom constrained by the bones and soft tissue structures of the joint. Similar to the other joints in the skeletal system, the main functions of hip and knee joints are to transfer the load from one bone to another, and to enable relative motion between the bones. The prosthetic hip and knee should also be able to provide these functions without failure, which is possibly attained by the implementation of an appropriate material with specific characteristics whose values approach those of natural biological materials (bone and tissue), for a given geometry.

The most paramount nonmechanical property for every material to be used in the human body is biocompatibility, which can be defined as the capability of a material to exist in contact with tissues without making an unacceptable degree of harm to the body. This property is used for screening out inappropriate materials. Furthermore, the ability of the material to integrate with the surrounding bone, known as osseointegration, is also important, especially for orthopedic materials where the implant should strongly be connected to the bone. The first mechanical property to be taken into account for load-bearing implants is adequate mechanical strength (yield or ultimate strength). When the implant is loaded in the same way as the natural bone, it seems rational that the material used can offer the same or even greater load-bearing capacity than the natural bone (compressive strength of compact bone is about 140 MPa). Furthermore, the fatigue endurance limit of the implant material is an important factor as well due to the repeated cyclic loads in the daily activities. A live healthy bone has self-healing characteristics to repair any microfractures and consequently possesses a great resistance to fatigue loading. Fatigue fracture is frequently observed with hip implants; however it is a very rare problem after total knee arthroplasty, particularly for the femoral component (Lemaire, 2010; Teoh, 2000). The other important property is the capability of the material to absorb energy during abnormal loading conditions (high-energy or high-rate loading situations such as a fall or jump). Possessing high-energy absorption and ductility is crucial to prevent brittle fracture of the implant under mechanical loading.

The other point is that bone is a living tissue with the ability to change its mineral density and structure in response to the loading environment (Wolff's law). If the stress experienced by the bone falls below its normal level, bone loss occurs, which might weaken the bone, deteriorate the implant-bone interface, and subsequently loosening and ultimate failure of the implant. This phenomenon is known as stress-shielding effect. In contrast, when the bone sustains higher stress values than its normal level, bone growth occurs. Adding the implant in the skeletal system usually shields the stresses from the bone and causes bone loss, which is associated with a large difference between elastic modulus of the implant material and the host bone. The elastic modulus of a compact bone is about 14 GPa in the longitudinal direction and about one-third of that in the radial direction. These values are modest compared to most engineering materials such as Co-Cr alloy, which has an elastic modulus of about 200 GPa. A lower elastic modulus, additionally, causes higher damping capacity, which can influence the absorption of impact energy and dampening of the peak stress between the bone and the prosthesis (Bahraminasab and Jahan, 2011).

Since both the hip and knee prostheses have multiple components, which are in contact with each other, wear can occur. The excessive occurrence of wear causes change both in the implant shape and probably the mechanical function, and in the biological response of the human body to the wear debris. Corrosion is another important issue which is unavoidable for metallic materials in the corrosive body fluid. The implants usually release nonbiocompatible metal ions leading to several health problems because the dissolved metal ions either can accumulate in the tissues adjacent to the implant, or they might be transported to other parts of the body. Corrosion can decrease both the life of the implant and human life. For polymeric materials, corrosion is not the main problem but leaching and absorption are important considerations (Kutz, 2002). In leaching phenomenon, small polymer chains, unreacted monomer molecules, or fillers may diffuse from the polymer component (acetabular cup or tibial insert) and disperse into the surrounding fluid, which possibly negatively influences the physiology and additionally reduces the density of the polymer component leading to adverse changes in the properties of the structure. Conversely, in the absorption process, water molecules, proteins, or lipids diffuse from the body fluid into the bulk of the polymer and distribute between the polymer molecules leading to a reduction in the mechanical strength and vulnerability to wear. Materials selection and material tailoring for hip and knee prostheses should therefore be done with respect to the above mentioned requirements to achieve acceptable performance, as described in the following sections.

7.3   Material selection for hip prosthesis

A hip prosthesis comprises of three main components: (1) femoral component (consisting of a stem/pin and head), (2) acetabular cup, and (3) acetabular interface. The femoral component is a rigid metal pin (manufactured as a precision machined forged cobalt chrome or titanium alloy but previously also in stainless steel, with either an integral ground and polished head or separately attached ceramic or metal ball head) that is implanted into the hollowed out shaft of the femur, replacing the natural femoral head. The hip socket (acetabulum) is inserted with a metallic or ceramic acetabular cup, which is fixed to the ilium. The acetabular interface is placed between the femoral component and the acetabular cup. This combination makes the articular surfaces for the hip joint and comes in a variety of material combinations (metal on polypropylene, ceramic on ceramic, and metal on metal) to reduce wear debris generated by friction. The pin and cup are usually fixed to the surrounding bone structure by adhesive bone cement and perform different functions. Fig. 7.1 shows the typical form and position of a hip joint prosthesis.

In this example, the material for the pin being considered has the requirements of tissue tolerance, corrosion and wear resistance, mechanical behavior (including tensile strength, fatigue strength, and relative toughness), elastic compatibility, weight, and cost. The objectives of the designer are to maximize the tissue tolerance, corrosion and wear resistance, and the mechanical behavior. However, the cost should be minimized, with target values for bone properties being elastic compatibility and weight. Table 7.1 shows the pin's candidate materials, criteria, objectives of the designer, and subjective weightings (Farag, 1997). C1, C2 are ordinal data or categorical data where there is a logical ordering to the categories,

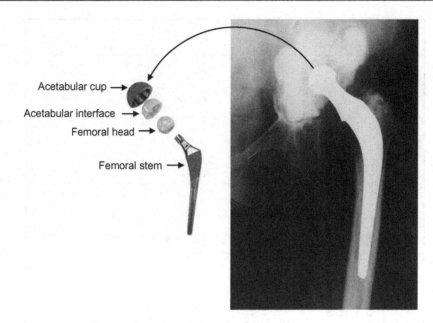

Acetabular cup →
Acetabular interface →
Femoral head →
Femoral stem →

Figure 7.1 Typical form and position of hip joint prosthesis.

that have been used for description of nonnumeric attributes; C3, C4, C7, and C8 are numeric attributes that represent absolute measure of material properties; and finally C5, C6, and C9 are ratio values. The correlation of criteria for materials selection of the hip joint prosthesis and final weighting of dependency are demonstrated in Table 7.2. Table 7.3 shows normalized values of hip joint prosthesis materials for all criteria. The subjective weightings are the same as those considered by Farag (1997). Table 7.4 shows objective weightings that have been extracted from Table 7.3. Table 7.5 is a summary of the weightings, including the objective weighting, weighting of dependency and finally subjective weighting. Also, the final weighting using different levels of importance for three types of weighting is shown in Table 7.5.

In this case, with regard to the type and number of criteria, the new versions of both TOPSIS (Jahan et al., 2012) and VIKOR (Jahan et al., 2011b) methods can fit as well for ranking the materials. Table 7.5 shows the ranking orders of materials using different weightings and ranking methods. When the data obtained using the comprehensive VIKOR, with $\lambda = 1$ that means only subjective weighting, were compared with the extended TOPSIS method, only Materials 5 and 8 changed places (rank 2 and 3). With $\lambda = 0.4$, Material 6 is the best in extended TOPSIS while with the same value of λ in the comprehensive VIKOR, Material 8 is the top ranked. However, most of the time for both methods, the top ranking belongs to Material 6. When tracking the ranking of Material 5, it shows that decreasing λ in extended TOPSIS causes a change in ranking of this material from 3 to 6 and in the

Table 7.1 Decision matrix for hip joint prosthesis materials selection

Objectives of design	Max.	Max.	Max.	Max.	Max.	Max.	Target value	Target value	Min.
Criteria	C1	C2	C3	C4	C5	C6	C7	C8	C9
	Tissue tolerance	Corrosion resistance	Tensile strength (MPa)	Fatigue strength (MPa)	Relative toughness	Relative wear resistance	Elastic modulus (GPa)	Specific gravity (g/cc)	Cost
Stainless steel 316	10	7	517	350	8	8	200	8	1
Stainless steel 317	9	7	630	415	10	8.5	200	8	1.1
Stainless steel 321	9	7	610	410	10	8	200	7.9	1.1
Stainless steel 347	9	7	650	430	10	8.4	200	8	1.2
Co-Cr alloys—Cast alloy (1)	10	9	655	425	2	10	238	8.3	3.7
Co-Cr alloys— Wrought alloy (2)	10	9	896	600	10	10	242	9.1	4
Unalloyed Titanium	8	10	550	315	7	8	110	4.5	1.7
Ti-6Al-4V	8	10	985	490	7	8.3	124	4.4	1.9
Composites (fabric reinforced)-Epoxy— 70% glass	7	7	680	200	3	7	22	2.1	3
Composites (fabric reinforced)-Epoxy— 63% carbon	7	7	560	170	3	7.5	56	1.6	10
Composites (fabric reinforced)-Epoxy— 62% aramid	7	7	430	130	3	7.5	29	1.4	5

Table 7.2 Correlation of criteria and weighting of dependency for a hip joint prosthesis

	C1	C2	C3	C4	C5	C6	C7	C8	C9	R_{jk}	W^c
C1	1.00	0.16	0.23	0.78	0.53	0.79	-0.97	-0.96	0.48	6.96	0.102
C2	0.16	1.00	0.57	0.48	-0.01	0.48	-0.18	-0.05	0.13	6.42	0.095
C3	0.23	0.57	1.00	0.73	0.29	0.46	-0.30	-0.21	0.16	6.07	0.089
C4	0.78	0.48	0.73	1.00	0.68	0.79	-0.85	-0.80	0.49	5.70	0.084
C5	0.53	-0.01	0.29	0.68	1.00	0.25	-0.62	-0.67	0.62	6.93	0.102
C6	0.79	0.48	0.46	0.79	0.25	1.00	-0.81	-0.76	0.13	6.66	0.098
C7	-0.97	-0.18	-0.30	-0.85	-0.62	-0.81	1.00	0.99	-0.47	11.21	0.165
C8	-0.96	-0.05	-0.21	-0.80	-0.67	-0.76	0.99	1.00	-0.51	10.97	0.162
C9	0.48	0.13	0.16	0.49	0.62	0.13	-0.47	-0.51	1.00	6.97	0.103

Table 7.3 Normalizing of hip joint prosthesis materials

Row	Materials	C1	C2	C3	C4	C5	C6	C7	C8	C9
1	Stainless steels 316	1.0000	0.0000	0.1568	0.4681	0.7500	0.3333	0.1842	0.2338	1.0000
2	Stainless steels 317	0.6667	0.0000	0.3604	0.6064	1.0000	0.5000	0.1842	0.2338	0.9889
3	Stainless steels 321	0.6667	0.0000	0.3243	0.5957	1.0000	0.3333	0.1842	0.2468	0.9889
4	Stainless steels 347	0.6667	0.0000	0.3964	0.6383	1.0000	0.4667	0.1842	0.2338	0.9778
5	Co-Cr alloys—Cast alloy (1)	1.0000	0.6667	0.4054	0.6277	0.0000	1.0000	0.0175	0.1948	0.7000
6	Co-Cr alloys—Wrought alloy (2)	1.0000	0.6667	0.8396	1.0000	1.0000	1.0000	0.0000	0.0909	0.6667
7	Unalloyed Titanium	0.3333	1.0000	0.2162	0.3936	0.6250	0.3333	0.5789	0.6883	0.9222
8	Ti-6Al-4V	0.3333	1.0000	1.0000	0.7660	0.6250	0.4333	0.5175	0.7013	0.9000
9	Composites (fabric reinforced)—Epoxy—70% glass	0.0000	0.0000	0.4505	0.1489	0.1250	0.0000	0.9649	1.0000	0.7778
10	Composites (fabric reinforced)—Epoxy—63% carbon	0.0000	0.0000	0.2342	0.0851	0.1250	0.1667	0.8158	0.9351	0.0000
11	Composites (fabric reinforced)—Epoxy—62% aramid	0.0000	0.0000	0.0000	0.0000	0.1250	0.1667	0.9342	0.9091	0.5556

Table 7.4 Objective weightings

	C1	C2	C3	C4	C5	C6	C7	C8	C9
Standard deviation	0.405	0.433	0.290	0.305	0.412	0.317	0.362	0.349	0.298
Objective weighting	0.128	0.137	0.092	0.096	0.130	0.100	0.114	0.110	0.094

Table 7.5 Final weightings using different effect of subjective, objective, and dependency weightings

	C1	C2	C3	C4	C5	C6	C7	C8	C9
Subjective weighting	0.200	0.200	0.080	0.120	0.080	0.080	0.080	0.080	0.080
Objective weighting	0.128	0.137	0.092	0.096	0.130	0.100	0.114	0.110	0.094
Dependency weighting	0.102	0.095	0.089	0.084	0.102	0.098	0.165	0.162	0.103
$W(\lambda=0,(1-\lambda)/2=0.5)$	0.115	0.116	0.090	0.090	0.116	0.099	0.140	0.136	0.098
$W(\lambda=0.2,(1-\lambda)/2=0.4)$	0.132	0.132	0.088	0.096	0.109	0.095	0.128	0.125	0.095
$W(\lambda=0.4,(1-\lambda)/2=0.3)$	0.149	0.149	0.086	0.102	0.102	0.091	0.116	0.114	0.091
$W(\lambda=0.6,(1-\lambda)/2=0.2)$	0.166	0.166	0.084	0.108	0.094	0.088	0.104	0.102	0.087
$W(\lambda=0.8,(1-\lambda)/2=0.1)$	0.183	0.183	0.082	0.114	0.087	0.084	0.092	0.091	0.084
$W(\lambda=1,(1-\lambda)/2=0)$	0.200	0.200	0.080	0.120	0.080	0.080	0.080	0.080	0.080

Table 7.6 Aggregated result and individual ranks of hip prosthesis materials generated by extended TOPSIS and comprehensive VIKOR with different weightings

Materials	Ranking with different values of λ												Aggregated ranking
	Extended TOPSIS						Comprehensive VIKOR						
	1	0.8	0.6	0.4	0.2	0	1	0.8	0.6	0.4	0.2	0	
1	5	5	5	5	7	9	5	5	5	7	8	8	5
2	7	7	7	7	6	5	7	7	7	6	6	6	7
3	8	8	8	8	8	7	8	8	8	8	7	7	8
4	6	6	6	6	5	4	6	6	6	5	5	5	6
5	3	3	4	4	4	6	2	2	2	4	4	10	4
6	1	1	1	2	2	3	1	1	1	1	3	3	1
7	4	4	3	3	3	2	4	4	4	3	2	2	3
8	2	2	2	1	1	1	3	3	3	2	1	1	2
9	9	9	9	9	9	8	9	9	9	9	9	4	9
10	11	11	11	11	11	11	11	11	11	11	11	11	11
11	10	10	10	10	10	10	10	10	10	10	10	9	10

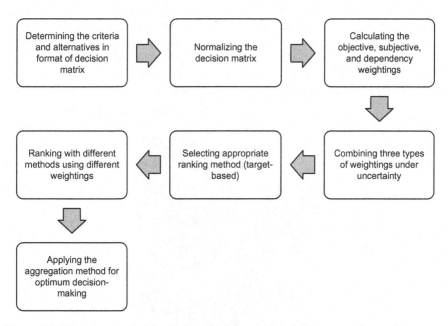

Figure 7.2 The stages applied for hip prosthesis materials selection example.

comprehensive VIKOR from 2 to 10. As a result, these rapid changes highlight the need for the aggregation technique (Jahan et al., 2011a) for the final optimum ranking that has been shown in Table 7.6. Fig. 7.2 shows how a combination of tools and techniques can be used to select the optimum material for the hip prosthesis.

7.4 Materials selection for knee prosthesis

A total knee prosthesis usually consists of three main components: (1) femoral
component, (2) tibial component (including tibial tray and tibial insert), and (3)
patellar component (also known as the kneecap). The femoral component replaces
the distal femur and the tibial tray is inserted into the proximal tibia. The tibial
insert is situated between the femoral and tibial components and the patellar component replaces the back of the patella. These two components, which are usually
made of polymers (eg, ultra-high molecular weight polyethylene), both articulate
against the prosthetic femur and attempt to mimic the natural knee constraints and
motion. The femoral component and the tibial tray, which are often made from
metallic materials (eg, Co-Cr alloys and Ti alloys) and sometimes from ceramics
(eg, alumina and zirconia), are typically cemented or pressed in to place. Fig. 7.3
shows the components of the knee prosthesis. This example ranks the candidate
materials for the femoral component of the knee prosthesis. In this example, metallic biomaterials (biologically compatible materials) that are currently used, and
newly developed metallic biomaterials that could potentially be used in the future
for the femoral component of knee joint implants are considered as candidate
materials (Bahraminasab and Jahan, 2011; Jahan and Edwards). The material property criteria considered here include tensile strength, Young's modulus, ductility,
corrosion resistance, wear resistance, and osseointegration ability. The metals
currently used are stainless steels, Co-Cr alloys, titanium, and titanium alloys. The
metals that could potentially be used in the future are NiTi shape memory alloys
(SMA) including dense and porous NiTi. The candidate metallic materials are
therefore (1) stainless steel L316 (annealed), (2) stainless steel L316 (cold worked),

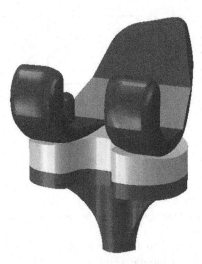

Figure 7.3 The main components of a total knee replacement (upper part: femoral
component; middle part: Tibial insert; lower part: Tibial component).

(3) Co-Cr alloys (wrought Co-Ni-Cr-Mo), (4) Co-Cr alloys (castable Co-Cr-Mo), (5) Ti alloys (pure Ti), (6) Ti alloys (Ti-6 Al-4V), (7) Ti-6Al-7Nb (IMI-367 wrought), (8) Ti-6Al-7Nb (Protasul-100 hot-forged), (9) NiTi shape memory alloy, and (10) porous NiTi shape memory alloy. Table 7.7 shows the properties of these materials $1-10$. This case study includes interval data, incomplete data, linguistic terms, and target criteria. The fuzzy conversion scales (Rao and Davim, 2008; Rao, 2006; Chen et al., 1992), systematically converts linguistic terms into their corresponding fuzzy numbers and used to assign the values of the attributes on a qualitative scale. An 11-point scale (Table 7.8) is used here for better understanding and representation of the qualitative attributes and converting linguistic terms into corresponding numbers (Table 7.9).

The application of the Interval target—based VIKOR (described previously in chapter: Multi-attribute decision-making for ranking of candidate materials) is illustrated in this case, step by step, as follows:

Step 1: Determining the most favorable values for all criteria.

$$T_j = (1.3, 1240, 16, 54, 0.96, 0.96, 0.96)$$

The target criteria are density and modulus of elasticity near to that of human bone, and for all the other properties the higher the better.

Step 2: Subjective weightings that have been calculated based on modified digital logic (MDL) (Dehghan-Manshadi et al., 2007). In the MDL method the scale of scores for the weighted factors is considered to be "1," "2," or, "3": "1" for attributes that are less important; "2" for attributes which are equally important; and "3" for attributes that are more important. More specifically, for each pair of attributes the question asked is, which attribute (material property) is more important for the desired outcome of the end product, property A or property B? The total number of possible decisions (pair-wise comparisons), N, is

$$N = \frac{n(n-1)}{2}$$

where n is the number of properties/criteria under consideration. After each combination is compared and assigned a binary score, the results are put into a matrix form as shown in Table 7.10, and the number of positive decisions for each attribute (ie, with a score of "1") is summed and a weighting factor, W_j is calculated and normalized such that

$$\sum W_j = 1; \quad j = 1, \dots, n$$

Step 3: Table 7.11 shows the required parameter for normalizing the materials (Table 7.12). Computing values $S_i^L, S_i^U, R_i^L, R_i^U, Q_i^L$ and Q_i^U with $\nu = 0.5$ is demonstrated in Table 7.13.

Step 4: Table 7.13 shows details of calculations. According to the guideline of comparing interval data (Table 7.8), Material 10 (porous NiTi shape memory alloy) has the minimum value of $Q_i([Q_i^L, Q_i^U])$. The ranking of other materials is shown in Table 7.14. When considering only one criterion such as wear resistance, Co-Cr alloy becomes the highest ranking in the currently used materials whereas it now has second ranking.

Table 7.7 **Properties of candidate materials for femur component materials selection**

Objectives	Target	Max.	Target	Max.	Max.	Max.	Max.
Material number	Density (g/cc)	Tensile strength (MPa)	Modulus of elasticity (GPa)	Elongation (%)	Corrosion resistance	Wear resistance	Osseointegration
1	8	517	200	40	High	Above average	Above average
2	8	862	200	12	High	Very high	Above average
3	9.13	896	240	10–30	Very high	Extremely high	High
4	8.3	655	240	10–30	Very high	Extremely high	High
5	4.5	550	100	54	Exceptionally high	Above average	Very high
6	4.43	985	112	12	Exceptionally high	High	Very high
7	4.52	≥900	105–120	≥10	Exceptionally high	High	Very high
8	4.52	1000–1100	110	10–15	Exceptionally high	High	Very high
9	6.50	≥1240	≥48	12	Extremely high	Exceptionally high	Average
10	<4.3	1000	15	12	Very high	Exceptionally high	Exceptionally high

Table 7.8 Value of materials selection factors in format of 11-point scale

Qualitative measure of materials selection factor	Assigned value
Exceptionally low	0.045
Extremely low	0.135
Very low	0.255
Low	0.335
Below average	0.410
Average	0.500
Above average	0.590
High	0.665
Very high	0.745
Extremely high	0.865
Exceptionally high	0.955

Table 7.9 Decision matrix for femur materials selection with quantitative data

Material number	Density (g/cc)	Tensile strength (MPa)	Modulus of elasticity (GPa)	Elongation (%)	Corrosion resistance	Wear resistance	Osseointegration
1	8	517	200	40	0.665	0.59	0.59
2	8	862	200	12	0.665	0.745	0.59
3	9.13	896	240	20	0.745	0.865	0.665
4	8.3	655	240	20	0.745	0.865	0.665
5	4.5	550	100	54	0.955	0.59	0.745
6	4.43	985	112	12	0.955	0.665	0.745
7	4.52	900	112.5	10	0.955	0.665	0.745
8	4.52	1050	110	12.5	0.955	0.665	0.745
9	6.5	1240	48	12	0.955	0.955	0.5
10	4.3	1000	15	12	0.745	0.955	0.955

Table 7.10 Determination of relative importance of criteria using MDL method

Criteria	Number of possible decisions [$N = n(n-1)/2$]																					Sum	Weighting
	1	2	3	4	5	6	7	8	9	10	11	12	13	14	15	16	17	18	19	20	21		
Density	1	1	1	1	1	1																6	0.07
Tensile Strength	3						1	2	1	1	1											9	0.11
Modulus of Elasticity		3					3					3	1	1	1							12	0.14
Elongation			3					2				1				1	1	1				9	0.11
Corrosion resistance				3					3				3			3			1	2		15	0.18
Wear resistance					3					3				3			3		3		2	17	0.20
Osseointegration						3					3				3			3		2	2	16	0.19

Table 7.11 Required parameter for normalization of femur materials

	Density (g/cc)	Tensile strength (MPa)	Modulus of elasticity (GPa)	Elongation (%)	Corrosion resistance	Wear resistance	Osseointegration
T_j	1.3	1240	16	54	0.955	0.955	0.955
$x_j^{U_{max}}$	9.13	1240	240	54	0.955	0.955	0.955
$x_j^{I_{min}}$	4.3	517	15	10	0.665	0.59	0.5

Table 7.12 Normalized data for femur materials

	$V^L_{i,C1}$	$V^U_{i,C1}$	$V^L_{i,C2}$	$V^U_{i,C2}$	$V^L_{i,C3}$	$V^U_{i,C3}$	$V^L_{i,C4}$	$V^U_{i,C4}$	$V^L_{i,C5}$	$V^U_{i,C5}$	$V^L_{i,C6}$	$V^U_{i,C6}$	$V^L_{i,C7}$	$V^U_{i,C7}$
1	0.8557	0.8557	1	1	0.8178	0.8178	0.3182	0.3182	1.0172	1.0172	1.0137	1.0137	0.8132	0.8132
2	0.8557	0.8557	0.5228	0.5228	0.8178	0.8178	0.9545	0.9545	1.0172	1.0172	0.589	0.589	0.8132	0.8132
3	1	1	0.4758	0.4758	0.9956	0.9956	1	0.5455	0.7414	0.7414	0.2603	0.2603	0.6484	0.6484
4	0.894	0.894	0.8091	0.8091	0.9956	0.9956	1	0.5455	0.7414	0.7414	0.2603	0.2603	0.6484	0.6484
5	0.4087	0.4087	0.9544	0.9544	0.3733	0.3733	0	0	0.0172	0.0172	1.0137	1.0137	0.4725	0.4725
6	0.3997	0.3997	0.3527	0.3527	0.4267	0.4267	0.9545	0.9545	0.0172	0.0172	0.8082	0.8082	0.4725	0.4725
7	0.4112	0.4112	0.4703	0.4703	0.3956	0.4622	1	1	0.0172	0.0172	0.8082	0.8082	0.4725	0.4725
8	0.4112	0.4112	0.332	0.1936	0.4178	0.4178	1	0.8864	0.0172	0.0172	0.8082	0.8082	0.4725	0.4725
9	0.6641	0.6641	0	0	0.1422	0.1422	0.9545	0.9545	0.0172	0.0172	0.0137	0.0137	1.011	1.011
10	0.3831	0.3831	0.332	0.332	0.0044	0.0044	0.9545	0.9545	0.7414	0.7414	0.0137	0.0137	0.011	0.011

Table 7.13 Interval numbers for S, R, and Q for femur materials

Material number	S^L	S^U	R^L	R^U	Q^L	Q^U
1	0.855	0.855	0.203	0.203	1.000	1.000
2	0.788	0.788	0.182	0.182	0.791	0.791
3	0.625	0.675	0.135	0.135	0.306	0.352
4	0.654	0.704	0.135	0.135	0.333	0.378
5	0.480	0.480	0.203	0.203	0.658	0.658
6	0.484	0.484	0.162	0.162	0.368	0.368
7	0.499	0.508	0.162	0.162	0.382	0.390
8	0.459	0.486	0.162	0.162	0.345	0.371
9	0.369	0.369	0.192	0.192	0.481	0.481
10	0.306	0.306	0.133	0.133	0.000	0.000

Table 7.14 Comparing ranking orders of candidate materials for femur component

Material number	Material	Interval target−based VIKOR
1	Stainless steel L316 (annealed)	10
2	Stainless steel L316 (cold worked)	9
3	Co-Cr alloys (wrought Co-Ni-Cr-Mo)	2
4	Co-Cr alloys (castable Co-Cr-Mo)	4
5	Ti alloys (pure Ti)	8
6	Ti alloys (Ti-6 Al-4V)	5
7	Ti-6Al-7Nb (IMI-367 wrought)	6
8	Ti-6Al-7Nb (Protasul-100 hot-forged)	3
9	NiTi shape memory alloy	7
10	Porous NiTi shape memory alloy	1

However, the materials above, particularly the top ranked, can fulfill the necessary mechanical and biological requirements, to some extent. The long-term success of knee implants, which is of major significance for young patients, focuses a need on the design of new materials with special properties not available in current materials. The following section recalls the example explained previously in chapter "Multiple objective decision-making for material and geometry design" and describes in detail the design and optimization of a functionally graded material (FGM) for the femoral component of the knee prosthesis.

7.5   FGM design for knee prosthesis

Aseptic loosening is one of the most challenges associated with knee prostheses, and it is caused by excessive wear in the prosthetic knee joint, stress shielding of the bone by prosthesis, and the development of soft tissue at the bone/implant interface due to relative motion of prosthesis. To address this problem via a material route, the design of a FGM can be a useful alternative. The current femoral components consist of a single material (usually a Co-Cr alloy), which essentially offers a uniform composition and structure. This component typically has a coating layer to better perform in terms of for example corrosion or osseointegration. Despite the benefit provided by the coating, the sudden changes of properties and the formation of a discrete boundary can be problematic.

To provide characteristics that avoid the leading causes of aseptic loosening, compositional changes should ideally be done continuously, from a material with high hardness and high modulus of elasticity at the articulation surface to a material with high bioactivity and low elastic modulus at bone/implant interface in a single integrated femoral component. Alumina-ceramic has proven to be a highly wear resistant material, particularly against the polyethylene insert (Oonishi et al., 2006).

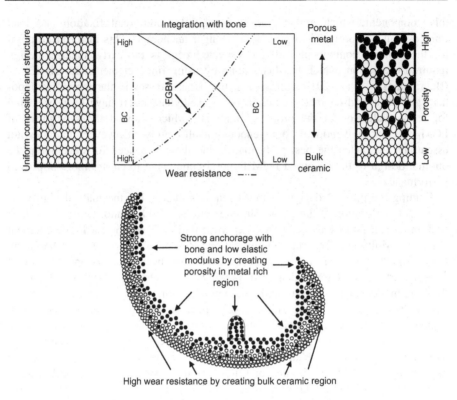

Figure 7.4 Proposed FGM and porous peg to overcome aseptic loosening.

Titanium has much lower elastic modulus (110 GPa) than that of alumina (350 GPa) and it is expected that this material can reduce the stress-shielding effect. Applying a gradient porous structure can further reduce the elastic modulus of the material and at the same time allow the bone cells to penetrate into the implant material. The maximum porosity should be at the bone-implant interface, and should gradually decrease to a negligible amount at the articulating surface. Hence, a porous alumina-titanium FGM can be promising for the femoral component to improve wear behavior, stress shielding, and relative micromotion simultaneously. Fig. 7.4 shows schematically the expected properties of the designed FGM. The properties of the materials are presented in the middle with the uniform material (bulky ceramic (BC)) on the left and the FGM on the right. As it can be seen in the figure, the wear resistance and the ability to connect with host bone are constant all through a uniform material. In the FGM component, the composition and structure are gradually changed from a ceramic with high wear resistant material at the bottom to a low modulus material with good bonding ability, by increasing the concentration of metallic material and amount of porosity at the top of the component. In addition to the main body of the femoral component, the pegs of this component, which are essentially located for additional fixation and load transfer, can be porous Ti to provide a stronger anchorage. This material proposal for the femoral component and the respected pegs is the early material design required evaluation, which has been done by using finite element analysis (FEA) (Bahraminasab et al., 2013). Stresses in the bone, stresses in the femoral component and pegs, and contact stresses in the polyethylene insert have been predicted for early evaluation of the proposed material, which showed the superiority of FGM over the conventional homogeneous metallic alloys. However, this does not essentially guarantee the best performance and there is a need for developing the optimal design of the proposed FGM femoral component to ensure more improved performance.

Finding the optimal design of FGM requires identifying the main design variables and exploration of their possible interactions, which cannot be directly clarified and quantified by conventional experimentation. Therefore, a Finite Element Analysis—Multiple Criteria Optimization—Response Surface Methodology (FEA—MCO—RSM) as a systematic approach has been used to optimize FGM design parameters (steps identified previously in chapter: Multiple objective decision-making for material and geometry design). The performance outputs of the FGM design are the stress distribution on the distal femur (mean and standard deviation), the contact characteristics of tibial insert (wear index), and micromotion at the implant/bone interface (mean and standard deviation). The five responses are related to the function of the material for reduction in aseptic loosening. The optimum conditions are to maximize the first performance output, and to minimize the second and third outputs, simultaneously, within the given constraints of the design variables. The determination of design variables is according to the "volume fraction" and "rule-of-mixtures" based on which the material properties of earlier design have been calculated. The compositions of alumina and titanium vary based on the relative length of (y/h), in relation to the material gradient k, where y is the distribution direction of each constituent and h is the height of the component (see Fig. 7.5). Factor k controls the variation in the volume fraction of titanium and alumina in the structure based on Eqs. (7.1) and (7.2). FGM has the highest amount of titanium for $k = 10$ and the richest content of alumina for $k = 0.1$. The variation in volume fraction of ceramic and metal, subsequently, alters the gradients in the Young's modulus and Poisson's ratio according to Eqs. (7.3)−(7.5).

Figure 7.5 A simple schematic of designed FGM, origin and y direction (black: metal, white: ceramic).

$$V_c = (y/h)^k \tag{7.1}$$

$$V_m = 1 - V_c \tag{7.2}$$

$$E = \frac{E_0(1-p)}{1 + p(5 + 8\nu)(37 - 8\nu)/\{8(1 + \nu)(23 + 8\nu)\}} \tag{7.3}$$

$$E_0 = E_c \left[\frac{E_c + (E_m - E_c)V_m^{2/3}}{E_c + (E_m - E_c)(V_m^{2/3} - V_m)} \right] \tag{7.4}$$

$$\nu = \nu_m V_m + \nu_c V_c \tag{7.5}$$

The volume fractions of metal is represented by V_m and the volume fractions of ceramic is represented by V_c, which are distributed over the y direction. The E and ν are Young's modulus and Poisson's ratio, respectively, at the different regions of the implant, E_0 is the equivalent elastic modulus at different regions of the implant without the porosity effect ($p = 0$), E_c is the Young's modulus of ceramic, E_m is the Young's modulus of metal and ν_c and ν_m are the Poisson ratios for ceramic and metal, respectively.

Furthermore, the porosity gradient in the FGM component can be adjusted by varying three arbitrary factors; A, n, and z based on Eqs. (7.6) and (7.7). For factor n, values other than 0 cause no porosity at the uppermost and lowermost surfaces of the FGM and the maximum porosity exists within the FGM. However, the femoral component needs to have the maximum porosity at the uppermost surface where it interfaces with the adjacent bone to provide good anchorage, avoiding micromotion. The value of n, therefore, needs to be zero here, in order to meet the design requirement. Factors A and z can vary to offer various amounts and gradients of porosity in the FGM implant. The changes in porosity alter the gradients of Young's modulus by Eq. (7.3). Meanwhile, the amount of porosity can affect the frictional behavior of the surface, which successively might influence the micromotion. Therefore, varying k, A, and z facilitates tailoring the property gradient, hence providing a means to optimize the performance brought by the FGM femoral component. Factor A changes between 0.4 and 0.7 to provide maximum porosity of 40−70% at the uppermost surface of the FGM. These values of porosity have been used to fabricate porous coated metals (Levine and Fabi, 2010). Factor z is adjusted between 0.5 and 2 because values higher and lower than these provide nonuniform variations in the structure and make sudden changes within the FGM, which is not desirable.

$$p = A(y/h)^n \{1 - (y/h)^z\} \tag{7.6}$$

$$\frac{((n+z)/n)^n}{1 - (n/(n+z))^z} \geq A \geq 0 \tag{7.7}$$

After determining the design variables of the FGM, Design of Experiments (DoE) is applied by using Central Composite Design (CCD) (Bahraminasab et al., 2014a). As described above, the three factors k, A, and z are the variables, which leads to a total of 15 design points including eight factorial (corner) points based on a 2^3 factorial design, one center point, and six axial (star) points at a face with distance of $\alpha = 1$ from the center point. Fig. 7.6 shows the design point based on CCD. Experimental runs $X_1 = \pm 1$, $X_2 = \pm 1$, $X_3 = \pm 1$ (X_1, X_2, and X_3 are factors k, A, and z, and $+1$ and -1 are upper and lower limits of the variables) are related to the factorial (corner) design points of the cube. Experimental run $X_1 = X_2 = X_3 = 0$ is associated with the center point of the design space. Because of applying computer-based simulation, only one single center point is taken into account due to no actual experimental error. Experimental runs $X_i = \pm 1$, $X_{i \neq j} = 0$ are indicative of the axial design points. Table 7.15 shows the corresponding uncoded values compared to the coded values.

By conducting the 15 experiments suggested by CCD and run by FEA, the 5 performance outputs (responses) are obtained and entered into a statistical software package (Minitab@14) to study the significance of the factors and their interactions and also to do the Analysis of Variance or ANOVA for each response. Table 7.16 shows the factors, their levels based on CCD and the responses obtained by FEA.

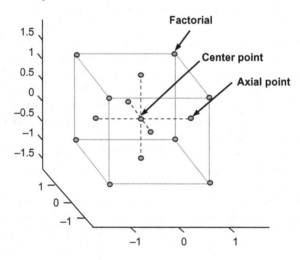

Figure 7.6 Graphical representation of a three-factor CCD.

Table 7.15 Coded and uncoded values for CCD test matrix

x_i	X_1 (k)	X_2 (A)	X_3 (z)
-1	0.1	0.4	0.5
0	5.05	0.55	1.25
1	10	0.7	2

Table 7.16 Factors (FGM design variables), levels, and responses

No.	Type	Factors			Responses				
		A:k	B:A	C:z	Mean stress (MPa)	Standard deviation of stress (MPa)	Mean micromotion (μm)	Standard deviation of micromotion (μm)	Wear index (MPa mm²)
1	Center	5.05	0.55	1.25	0.979	0.606	2.719	4.234	2906.963
2	Axial	5.05	0.55	2	0.980	0.606	2.747	4.252	2906.938
3	Axial	5.05	0.4	1.25	0.975	0.604	2.753	4.282	2907.020
4	Axial	5.05	0.55	0.5	0.976	0.605	2.695	4.201	2907.006
5	Axial	0.1	0.55	1.25	0.971	0.603	2.642	4.164	2907.130
6	Axial	5.05	0.7	1.25	0.985	0.608	2.716	4.207	2906.895
7	Axial	10	0.55	1.25	0.979	0.606	2.721	4.237	2906.948
8	Fact	10	0.4	0.5	0.974	0.604	2.741	4.284	2907.034
9	Fact	0.1	0.7	2	0.978	0.605	2.690	4.289	2907.019
10	Fact	0.1	0.4	2	0.969	0.602	2.691	4.235	2907.164
11	Fact	10	0.4	2	0.976	0.605	2.773	4.300	2906.988
12	Fact	0.1	0.4	0.5	0.967	0.601	2.666	4.216	2907.223
13	Fact	10	0.7	2	0.988	0.609	2.749	4.259	2906.847
14	Fact	10	0.7	0.5	0.982	0.607	2.690	4.230	2906.939
15	Fact	0.1	0.7	0.5	0.974	0.603	2.600	4.139	2907.140

It should be noted that checking the ANOVA assumptions, including normal distribution of residuals (the points should follow a straight line in normal probability plot), constant variance of residuals (the points should have a random pattern), and the consistency of a run with other runs via residuals, is a prerequisite of any interpretation. For analysis of the ANOVA model, the P-value of model is determinant; when the P-value of model is less than the significance level (α), which is usually set to be 0.05, then the model is significant. In the model (source) analysis, the sum of squares (SS) and mean sum of squares (MS) are calculated to evaluate the responses. F is the ratio of MS and MS_{Error} and is compared with F distribution tables used in statistics. The P-value signifies the probability of the hypothesis that a factor or treatment combination is not significant. The values of P are obtained from F distribution tables. In the analysis of terms, the coefficients (Coef) are used to build the mathematical model. The T is a test statistic with a student's t distribution and the P is associated with that test statistic. P-value $= 1$ specifies that a factor is least significant and P-value $= 0$ indicates the maximum significance. In the significant model, the influential factors (significant terms) are those have the P-value of less than $\alpha = 0.1$ which means the probability of the hypothesis that an effect (either main effect or second-order interaction) is significant is 90% true. The coefficients in the ANOVA table are used in the regression model for each response in order to do the multi-objective optimization process. Once the coefficients are acquired, usually, the adjusted R-squared (R-Sq$_{adj}$) is checked for goodness-of-fit (closer to 1 is better). More information about statistical analysis can be found in Montgomery and Runger (2010). Fig. 7.7 demonstrates the required steps and considerations used in applying RSM for design optimization.

In this example, all the responses, including mean and standard deviation of stresses in the bone, mean and standard deviation of micromotion, and wear index are identified to be significant. The ANOVA table for the second-order (full quadratic) uncoded model of mean stress is presented in Table 7.17. The significance of each estimate of regression parameter (term) is checked, to discard any noncontributing terms to the response (mean stress of bone) from the RSM full quadratic model. Therefore, considering the P-values in Table 7.17, no parameters are removed from the full mean stress model. Furthermore, Table 7.17 shows that the full second-order model is significant because the P-values of linear, square and interaction effects are all less than the threshold (0.05). The adjusted R^2 is 99.5%, which indicates 99.5% of variation in mean stress, and is described by the following full quadratic model (Eq. (7.8)):

$$
\begin{aligned}
\text{Mean of stress} = {} & 0.9731 + 0.0017 \times k - 0.0395 \times A + 0.0002 \times z \\
& - 0.0001 \times k^2 + 0.0526 \times A^2 - 0.0009 \times z^2 + 0.0006 \times k \times A \\
& + 0.0001 \times k \times z + 0.0076 \times A \times z
\end{aligned}
\tag{7.8}
$$

The main and interaction effects of factors are also evaluated. The main effects of the three factors are significant (see the related P-values in Table 7.17). Furthermore, the factors k, A, and z all interacted with each other. The interaction

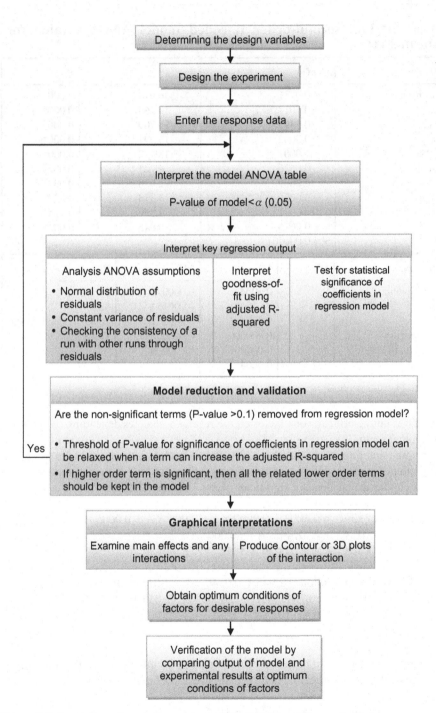

Figure 7.7 An outline of required steps in applying RSM for design optimization.

Table 7.17 Full second-order uncoded model ANOVA table for mean stress

Term	Coef	T	P
Constant	0.9731	4536.974	0.000[a]
k	0.0017	30.264	0.000[a]
A	−0.0395	36.709	0.000[a]
z	0.0002	14.991	0.000[a]
$k*k$	−0.0001	−13.172	0.000[a]
$A*A$	0.0526	4.725	0.005[a]
$z*z$	−0.0009	−2.038	0.097[a]
$k*A$	0.0006	3.378	0.020[a]
$k*z$	0.0001	2.356	0.065[a]
$A*z$	0.0076	6.045	0.002[a]
	$S = 0.0004013$	$R^2 = 99.8\%$	$R^2_{adj} = 99.5\%$

Source	DF	SS	MS	F	P
Regression	9	0.000444	0.000049	306.03	0.000[a]
Linear	3	0.000401	0.000134	829.39	0.000[a]
Square	3	0.000034	0.000011	70.85	0.000[a]
Interaction	3	0.000009	0.000003	17.83	0.004[a]
Residual error	5	0.000001	0.000000		
Total	14	0.000444			

[a]Denotes the significant factors.

occurs when the effect of one factor on the response changes depending on the level of another factor.

Similar interpretations, as above, for the second and third responses (standard deviation of stress and the mean micromotion respectively) show that the following full second-order models are significant (Eqs. (7.9) and (7.10)):

$$
\begin{aligned}
\text{STDV of stress} = {} & 0.6007 + 0.0009 \times k - 0.008 \times A + 0.0005 \times z \\
& - 0.0001 \times k^2 + 0.0144 \times A^2 - 0.0004 \times z^2 \\
& + 0.0001 \times k \times A + 0.0026 \times A \times z
\end{aligned} \tag{7.9}
$$

$$
\begin{aligned}
\text{Mean of micromotion} = {} & 2.945 + 0.0246 \times k - 1.0151 \times A - 0.0281 \times z \\
& - 0.0015 \times k^2 + 0.7047 \times A^2 + 0.0041 \times z^2 - 0.0014 \times k \times A \\
& - 0.0008 \times k \times z + 0.1020 \times A \times z
\end{aligned}
$$

$$\tag{7.10}$$

However, for standard deviation of micromotion (fourth response), the full second-order model is not significant, thus, the first-order model is considered (Eq. (7.11)).

$$
\text{STDV of micromotion} = 4.2348 + 0.0054 \times k - 0.01287 \times A + 0.0353 \times z
$$

$$\tag{7.11}$$

The reduced second-order model of wear index is given in Eq. (7.12), in which $A*A$ and $k*A$ are removed due to P-values greater than 0.1.

$$\text{Wear index} = 2907.3636 - 0.0515 \times k - 0.2440 \times A - 0.0291 \times z \\ - 0.0031 \times k^2 + 0.0146 \times z^2 + 0.0014 \times k \times z - 0.1194 \times A \times z \tag{7.12}$$

The next step after building the mathematical models is the optimization which is carried out by a "weighted metric" method (L_p method) (Noorossana and Ardakani, 2009). This approach combines several objectives into a one single objective. The weighted L_p distance is considered as a measure of any solution x from the ideal solution $f(X^{ideal})$, which can be minimized (see Eq. (7.13)):

$$L_p = \left\{ \sum_{j=1}^{k} w_j (|f_j(X^{ideal}) - f_j(X)|)^P \right\}^{1/P} \tag{7.13}$$

In the above equation, $f(X)$ is the objective function (response), while j can be 1, 2, 3..., k in which k is the number of objective functions. The term w_j denotes a weighting (nonnegative) that the decision-maker assign to each objective function and P signifies the importance of each objective function deviation from its ideal value. Eq. (7.13) is typically used where all objective functions are in the same scale. For the occasion in which $f(X)$s are in different scales, such as the responses in this case study, each objective function can become scale-less by Eq. (7.14):

$$L_p = \left(\sum_{j=1}^{k} w_j \left(\left| \frac{f_j(X^{ideal}) - f_j(X)}{f_j(X^{max}) - f_j(X^{min})} \right| \right)^P \right)^{1/P} \tag{7.14}$$

The ideal and nonideal values for the responses are obtained by either minimizing or maximizing each response individually (single objective). These values and the corresponding values for the three factors (k, A, and z) are shown in Tables 7.18 and 7.19 in which the ideal and nonideal responses are underlined. The ideal and nonideal values are used in Eq. (7.14) to obtain the optimum solution for the multi-objective problem. Here, identical weightings are assigned to all the objective functions and the optimal values of factors are attained for different P values as demonstrated in Table 7.20.

Varying the value of P has negligible effect on the optimal values of the factors and the predicted responses. To ascertain the optimal values, computational experimentation at the optimum point is done using FEA. Table 7.21 shows the comparison between the predicted response values estimated by the mathematical model and by FEA at the optimum point. It can be seen that the differences are extremely small, demonstrating an acceptable correlation between the predictions of the mathematical model and the results of the FEA.

Table 7.18 The ideal values of objective functions and the related values of factors

	Mean stress (MPa)	Standard deviation of stress (MPa)	Mean micromotion (µm)	Standard deviation of micromotion (µm)	Wear index (MPa mm^2)	k	A	Z
Mean stress (MPa)	0.991864	0.604896	2.752933	4.350391	2906.228840	10	0.7	2
Standard deviation of stress (MPa)	0.967309	0.600567	2.661436	4.247842	2907.226109	0.1	0.4	0.5
Mean micromotion (µm)	0.973386	0.603102	2.604538	4.244185	2907.139813	0.1	0.684	0.5
Standard deviation of micromotion (µm)	0.973975	0.603312	2.604715	4.243981	2907.134999	0.1	0.7	0.5
Wear index (MPa mm^2)	0.991864	0.604896	2.752933	4.350391	2906.228840	10	0.7	2

Table 7.19 The nonideal values of objective functions and the related values of factors

	Mean stress (MPa)	Standard deviation of stress (MPa)	Mean micromotion (µm)	Standard deviation of micromotion (µm)	Wear index (MPa mm^2)	k	A	z
Mean stress (MPa)	0.967309	0.604115	2.624752	4.249941	2907.085886	0.1	0.4	0.5
Standard deviation of stress (MPa)	0.98756375	0.607548	2.754246	4.322581	2906.716725	4.85	0.7	2
Mean micromotion (µm)	0.979008	0.602720	2.777438	4.340644	2906.632958	7.48	0.4	2
Standard deviation of micromotion (µm)	0.979996	0.600684	2.767912	4.354252	2906.373680	10	0.4	2
Wear index (MPa mm^2)	0.967309	0.600567	2.661436	4.247842	2907.226109	0.1	0.4	0.5

Table 7.20 **Variation in the optimum values of factors by altering P**

P	k	A	z
1	10	0.7	0.5
2	10	0.685133	0.5
3	10	0.685099	0.5
4	10	0.686607	0.5
5	10	0.688545	0.5

Table 7.21 **Comparison of predicted values by model and FEA**

Responses	Predicted values by model	Predicted values by FEA	Errors (%)
Mean stress (MPa)	0.985459	0.981581	−0.40
Standard deviation of stress (MPa)	0.602916	0.606831	0.65
Mean micromotion (μm)	2.684608	2.689573	0.18
Standard deviation of micromotion (μm)	4.297441	4.22984	−1.60
Wear index (MPa mm^2)	2906.322	2906.939	0.02

7.6 Materials selection and design optimization— location pegs for knee prosthesis

In this example, the optimized FGM femoral component with porous pegs, as explained in the previous example, and standard Co-Cr alloy femoral component, are used to perform the optimization of the performance with respect to a flexible geometrical design (location pegs) and to rank the optimal designs based on the materials used (various amount of porosity) (Bahraminasab et al., 2014b). Again, the aim is to reduce the primary causes of aseptic loosening (stress shielding, wear, and micromotion) simultaneously. To achieve this goal, three main steps are taken: (1) applying a combination of DoE and FEA to analyze the interactions, and to clarify the direction for improvement in goals regarding design variables, (2) obtaining the primary optimized peg geometry for each material by multi-objective optimization using *Design-Expert* software, and (3) building a database of promising designs and selecting the final design using the multi-attribute decision-making technique. Fig. 7.8 shows the steps used in the design methodology.

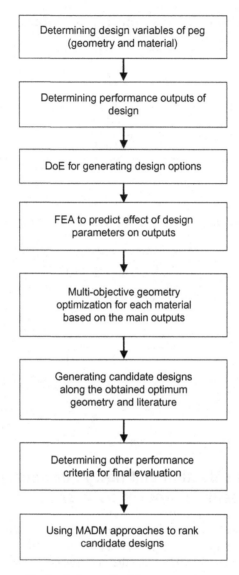

Figure 7.8 The steps used in the design methodology.

Three design variables are considered, including two numerical factors (peg length (L) and peg diameter (D)) and one categorical factor (femoral component/peg material). Three objective functions (performance outputs) are also taken into account, including weighted mean and standard deviation of stresses within the femoral bone, and mean stress at the location pegs. A central composite rotatable experimental design (CCRD) is applied, in this case study, to plan the

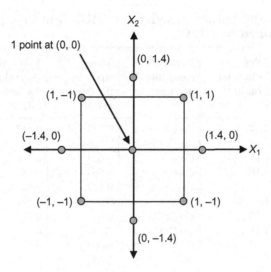

Figure 7.9 Graphical representation of central composite rotatable design with two numerical factors.

Table 7.22 Coded and uncoded values for CCRD in peg design analysis

Design points	Peg length (X₁) mm	Peg diameter (X₂) mm	Femoral component/ peg material (X₃)
−1.4	7.34	2.75	—
−1	9	4	Co-Cr/Co-Cr
0	13	7	—
1	17	10	FGM/40% porous Ti
1.4	18.66	11.24	—

computational experiments for peg geometrical design optimization and to find the relationships between the peg design parameters and the responses, as well. The CCRD used here consists of two combinations of 9 design points as shown in Fig. 7.9, leading to 18 experiments. The ranges and levels of variables investigated are shown in Table 7.22.

Afterward, the responses of 18 runs are predicted using FEA and entered into the *Design-Expert* software to use its optimization capabilities to perform the multi-objective optimization of the peg design. Table 7.23 shows the design points based on CCRD and the performance outputs obtained by FEA.

The *Design-Expert* software searches the design space and offers the optimization solution according to a desirability function. The overall desirability function

Table 7.23 **Design points based on CCRD and the performance outputs obtained by FEA**

No.	Peg length (mm)	Peg diameter (mm)	Femoral component/ peg material	Mean stress in the femur (MPa)	STDV of stresses in the femur (MPa)	Mean stress of peg (MPa)
1	9	4	Co-Cr/Co-Cr	0.623	0.153	8.96
2	17	4	Co-Cr/Co-Cr	0.626	0.183	15.96
3	9	10	Co-Cr/Co-Cr	0.607	0.16	5.86
4	17	10	Co-Cr/Co-Cr	0.568	0.155	6.87
5	7.34	7	Co-Cr/Co-Cr	0.622	0.154	6.29
6	18.66	7	Co-Cr/Co-Cr	0.595	0.171	11.45
7	13	2.76	Co-Cr/Co-Cr	0.625	0.162	17.87
8	13	11.24	Co-Cr/Co-Cr	0.581	0.162	5.7
9	13	7	Co-Cr/Co-Cr	0.603	0.161	8.65
10	9	4	FGM/40% porous Ti	0.658	0.158	6.65
11	17	4	FGM/40% porous Ti	0.66	0.188	10.97
12	9	10	FGM/40% porous Ti	0.639	0.165	3.71
13	17	10	FGM/40% porous Ti	0.6	0.158	5.51
14	7.34	7	FGM/40% porous Ti	0.654	0.16	3.78
15	18.66	7	FGM/40% porous Ti	0.626	0.166	8.00
16	13	2.76	FGM/40% porous Ti	0.66	0.167	12.41
17	13	11.24	FGM/40% porous Ti	0.612	0.165	4.33
18	13	7	FGM/40% porous Ti	0.637	0.165	6.32

concurrently considers all the responses and transforms multiple responses into a single response through mathematical transformations. The desirability function contains the transformation of each estimated response to a desirability value of between 0 and 1 and then selects the design variables in such a way as to maximize the overall desirability of all responses. The overall desirability is calculated by multiplication of individual desirability. More information and details of mathematical methods are available in Pasandideh and Niaki (2006), Karande et al (2012), Del Castillo (1996), Del Castillo and Montgomery (1993), and Derringer and Suich (1980). In the *Design-Expert* software, the optimization process starts by determining a goal (maximum, minimum, or target) for each response, and applying constraints (lower and upper limits). In this example, the goal is to maximize the mean stress in the bone, while minimizing the STDV of stresses in the bone, and mean stress in the pegs. The ranges of variations for peg length and peg diameter are defined to be 9–17 and 4–10 mm, respectively. After that, the weightings and importance of each response are defined in which the weightings can be between 0.1 and 10 and the relative importance of one response versus another response, can vary from the lowest importance (+) to the most important (+++++). In this example, the weightings and importance of responses are considered to be equal for all design objectives, with a weighting of 1 assigned and importance of responses equal to +++. It is worth noting that a low weighting (close to 0.10) allows more solutions not quite meeting the optimal goal, and a high weighting (close to 10) searches for a solution near to or beyond the stated goal. From a practical perspective, leaving the weightings at 1.0 is a good place to begin.

In this example, the conflict of objectives is the main reason for the multi-objective optimization method. This is shown in Table 7.24, where maximizing the first response increases the second and third responses, which is not desirable. Similar to the previous example, the ideal solutions are obtained by optimizing a single objective, merely, in the design space. The results of optimization provided by *Design-Expert* software shows that for the FGM femoral component with 40% porous Ti pegs, the optimum values of length and diameter are 9 and 4.5 mm, respectively, with a desirability of 0.881. Furthermore, the point prediction (prediction of responses at the optimum point) of *Design-Expert* software for mean stress of femoral bone, STDV of stresses in femur, and mean stress of peg are 0.655, 0.157, and 6.24, respectively, which are slightly different from those estimated by FEA. The errors are 0%, 1.88%, and 0.32%, respectively, for the mean stress in the femur, STDV of stresses in the femur, and the mean stress of the peg which obviously reveal the agreement between the predicted values of the responses and the FEA results. Additionally, the *Design-Expert* software provides an optimal geometry for Co-Cr alloy with less desirability (0.717) in which the suggested optimum length and diameter are 9 and 6 mm, respectively.

The optimal geometry of the peg for each material obtained here, with different combinations of peg materials, is used to generate conceptual candidate designs for final selection. Bulk Ti, 40% porous Ti, and 60% porous Ti material properties are assigned for the FGM femoral component pegs to see the influence of porosity in the final ranking. Meanwhile, the FEA shows that the values of stress are quite high in the peg root therefore, this zone is critical because of the transition from the main body of the component to the peg. When the cross-section becomes porous, the high stress level in the peg root is even worse. This aspect is important due to the adverse effect on the component, not the stress transition into the bone, which may cause premature fracture of the peg via the root. This might bring the idea of a blend or fillet radius being introduced at the base of the peg (conical pegs). In addition to this, a four-peg design is also considered according to the suggestion in the literature. Fig. 7.10 shows the candidate designs used for the final ranking.

Table 7.24 Ideal solutions (targets) based on single-objective optimization

Objective	Design parameters			Mean stress in femur (MPa)	STDV of stresses in femur (MPa)	Mean stress in peg (MPa)	Desirability
	Peg length (mm)	Peg diameter (mm)	Femoral/peg material				
Maximizing mean stress in femur	17	4	FGM/porous Ti	*0.658*	0.182	12.01	0.973
	17	4	Co-Cr/Co-Cr	*0.624*	0.180	16.49	0.606
Minimizing STDV of stresses in femur	9.13	4.63	Co-Cr/Co-Cr	0.620	*0.151*	9.66	1.000
	9	4	FGM/porous Ti	0.656	*0.156*	6.86	0.912
Minimizing mean stress in peg	9	8.88	FGM/porous Ti	0.643	0.165	*3.87*	0.989
	9	10	Co-Cr/Co-Cr	0.608	0.162	*5.31*	0.887

Bold: The predicated optimum value of the objective specified in the first column for each material. Underline: The best predicted value between the two materials for the objective specified in the first column.

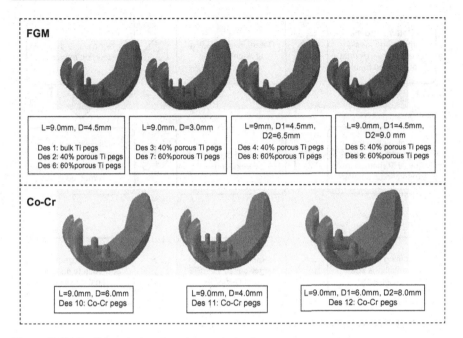

Figure 7.10 Candidate designs based on optimized peg geometry.

These conceptual candidate designs are modeled and analyzed by FEA, then based on the responses obtained MADM techniques are used for ranking. The design problem in this example includes target-based criteria, thus comprehensive VIKOR (C-VIKOR) (Jahan et al., 2011b) and extended TOPSIS methods (Jahan et al., 2012) are applied to rank the candidate designs. Dependency weightings between criteria, objective weightings and subjective weightings are considered (described previously in chapter: Multi-attribute decision-making for ranking of candidate materials). The analytic hierarchy process (AHP) method is used for extracting the subjective weightings using *Design-Expert* software to simplify pairwise comparison of criteria. Sensitivity analysis is carried out on the ranking results through various weightings obtained by combining three types of weightings (subjective, objective, and dependency) under uncertainty of their importance. Fig. 7.11 shows the steps considered in this example for design selection of the femoral component's peg.

The criteria used for the ranking stage includes the mean stress of femur, STDV of stress in femur, first quartile of stresses in femur, mean stress of peg, maximum over yield stress of peg, mass, and osseointegration. Table 7.25 presents the details of candidate designs, the criteria used, and the objectives. A higher level of mean stress in the bone is desired to diminish the stress-shielding effect and the consequential bone loss, but for STDV of stress in bone lower values are preferable due to less nonuniform stress distribution in the bone. For these criteria, ideal solutions obtained in Table 7.24 are used as target values. In addition to this, minimum stress

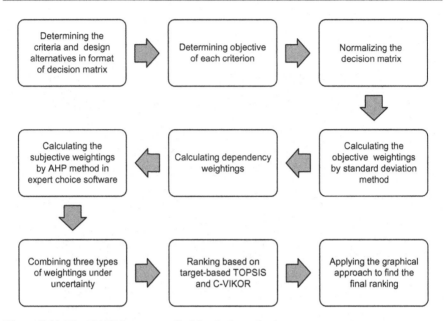

Figure 7.11 The MADM stages applied for design selection.

is also important in the bone loss initiation. However, it appears that the first quartile is more reliable than the minimum stress when an extreme observation exists (perhaps because of a single point/node). In descriptive statistics, the quartiles of a ranked set of data values are defined as the three points dividing the data set into four identical subsets, each subset is a quarter of the data in which the first quartile signifies a level that 75% of the data are higher than this level. Therefore, higher values of this criterion are preferred. Target values are also taken into account for mean stress of peg and mass. The mass in Table 7.25 is the summation of femoral component and the cut femur mass, and the target value is the mass of femoral bone before resection. The criterion maximum over yield stress of peg is considered to attain stress transition to the adjacent bone without over compromising the peg strength. The yield strength of bulk Ti (commercially pure titanium, grade 4) can get to 650 MPa which results in the yield strength of 90 and 49 MPa, respectively, for 40% and 60% porous Ti, by applying the equation of Gibson and Ashby (1999). Furthermore, the yield strength of Co-Cr alloy is considered to be 450 MPa. Additionally, osseointegration is considered as a criterion for anchorage to the bone.

In order to manipulate quantitative data only, an 11-point scale, similar to the second example of this chapter (Section 7.4) is used. This scale systematically translates linguistic term of osseointegration ability in Table 7.25 to their corresponding fuzzy numbers (Table 7.8). First the decision-making matrix is normalized with a target-based normalization technique (described previously in chapter: Multi-attribute decision-making for ranking of candidate materials), then objective and dependency weightings are calculated as presented in Table 7.26. For the subjective weightings of the design criteria, a pair-wise comparison method

Table 7.25 Conceptual candidate design of femoral component

Des No.	Main body material	Peg material	Peg geometry	Target (0.658) A	Target (0.151) B	Target (3.87) C	Min. D	Target (95.4) E	Max. F	Max. G
1	FGM	bulk Ti	L = 9, D = 4.5 mm	0.653	0.160	7.567	0.040	167	0.535	Average
2		40% porous Ti	L = 9, D = 4.5 mm	0.655	0.160	6.262	0.197	166	0.535	High
3		40% porous Ti	L = 9, D = 3 mm (4 pegs)	0.646	0.165	7.860	0.424	166	0.524	High
4		40% porous Ti	L = 9, D1 = 4.5, D2 = 6.5 mm (conical peg)	0.653	0.162	5.487	0.142	166	0.530	High
5		40% porous Ti	L = 9, D1 = 4.5, D2 = 9 mm (conical peg)	0.651	0.163	4.747	0.109	167	0.529	High
6		60% porous Ti	L = 9, D = 4.5 mm	0.656	0.160	5.176	0.239	166	0.537	Very high
7		60% porous Ti	L = 9, D = 3 mm (4 pegs)	0.649	0.164	6.480	0.464	166	0.528	Very high
8		60% porous Ti	L = 9, D1 = 4.5, D2 = 6.5 mm (conical peg)	0.654	0.162	4.510	0.190	166	0.531	Very high
9		60% porous Ti	L = 9, D1 = 4.5, D2 = 9 mm (conical peg)	0.653	0.163	3.842	0.137	166	0.531	Very high
10	Co-Cr	Co-Cr	L = 9, D = 6 mm	0.620	0.155	6.613	0.045	317	0.499	Low
11		Co-Cr	L = 9, D = 4 mm (4 pegs)	0.606	0.161	9.546	0.077	316	0.481	Low
12		Co-Cr	L = 9, D1 = 6, D2 = 8 mm (conical peg)	0.617	0.157	7.307	0.058	318	0.501	Low

A, mean stress in the femur (MPa); B, STDV of stresses in the femur (MPa); C, mean stress in the pegs (MPa); D, maximum over yield stress of peg material; E, mass; F, first quartile of stresses in the femur (MPa); G, osseointegration ability.

Table 7.26 **The types of weightings and result of their combination under uncertainty in importance of each weighting**

Design selection criteria	A	B	C	D	E	F	G
Subjective weighting (w_j^s)	0.245	0.118	0.077	0.140	0.065	0.119	0.236
Objective weighting (w_j^o)	0.155	0.095	0.129	0.150	0.139	0.144	0.189
Dependency weighting (w_j^c)	0.102	0.236	0.108	0.220	0.114	0.101	0.119
Final weightings for combination under uncertainty in importance of different types of weighting							
$W(\lambda = 0)$	0.128	0.165	0.118	0.185	0.127	0.123	0.154
$W(\lambda = 0.2)$	0.152	0.156	0.110	0.176	0.114	0.122	0.170
$W(\lambda = 0.4)$	0.175	0.146	0.102	0.167	0.102	0.121	0.187
$W(\lambda = 0.6)$	0.198	0.137	0.094	0.158	0.090	0.120	0.203
$W(\lambda = 0.8)$	0.222	0.127	0.085	0.149	0.077	0.120	0.220
$W(\lambda = 1)$	0.245	0.118	0.077	0.140	0.065	0.119	0.236

A, mean stress in the femur (MPa); B, STDV of stresses in the femur (MPa); C, mean stress in the pegs (MPa);
D, maximum over yield stress of peg material; E, mass; F, first quartile of stresses in the femur (MPa);
G, osseointegration ability.

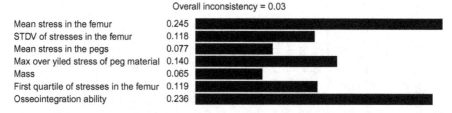

Synthesis with respect to: goal: Peg design selection

Overall inconsistency = 0.03

Mean stress in the femur	0.245
STDV of stresses in the femur	0.118
Mean stress in the pegs	0.077
Max over yiled stress of peg material	0.140
Mass	0.065
First quartile of stresses in the femur	0.119
Osseointegration ability	0.236

Figure 7.12 Overall results of the subjective weighting by AHP in *Design-Expert* software.

introduced in the AHP method is used and the final subjective weightings and the inconsistency ratio are shown in Fig. 7.12. The inconsistency ratio shows the degree of instability in the designer's judgments, which is recommended to have the value of less than 0.1 for a reliable decision.

For the sensitivity of design ranking over the design criteria, the final weightings are combined through Eq. (7.15). The terms w_j^o, w_j^s, and w_j^c denote the objective, subjective, and dependency weightings (correlation effects) correspondingly, and $0 \leq \lambda \leq 1$.

$$W_j = w_j^s \lambda + w_j^o \frac{(1 - \lambda)}{2} + w_j^c \frac{(1 - \lambda)}{2} \quad j = 1, 2, 3, \ldots, 7 \tag{7.15}$$

The ranking of designs using the extended TOPSIS method (a method for risk avoidance designers) are shown in Fig. 7.13. Based on the results obtained, the

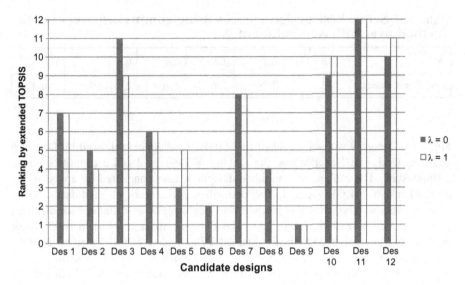

Figure 7.13 Comparative ranking of candidate designs for $\lambda = 0$ and 1 using extended TOPSIS method.

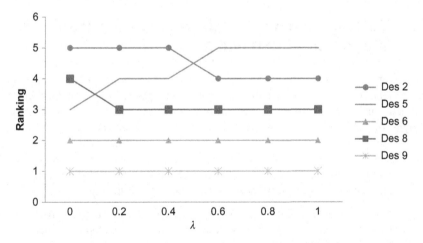

Figure 7.14 Stability of design's rankings for different values of λ using extended TOPSIS method.

FGM femoral component with either porous Ti or bulk Ti pegs are generally better ranked than the femoral component with fully Co-Cr (ie, Designs 10, 11, and 12). Furthermore, it is found that the four-peg design options (ie, Designs 3, 7, and 11) are not promising designs compared to the other designs. In contrast it seems that conical peg designs are high potential design candidates. Different values of λ are tested to ensure the ranking order of candidate designs. The stability of top rank designs over different values of λ is shown in Fig. 7.14. As it can be seen in

Table 7.27 Set of best design choices using comprehensive VIKOR method over different values of λ

λ	0	0.2	0.4	0.6	0.8	1
Set of best choices	2, 6	2, 6	2, 6	2, 6	2, 6, 8, 9	2, 6, 8, 9

Fig. 7.14, Design 9 is always ranked first (100% confidence) followed by Design 6 and then Design 8. Nevertheless, the ranking of Designs 2 and 5 cannot be easily distinguished. The ranking of candidate designs is also conducted by applying the comprehensive VIKOR method. The performance outcomes of candidate alternatives are very similar, thus this method provides a set of best choices for different values of λ as shown in Table 7.27. The outcome from comprehensive VIKOR method in conjunction with those obtained by extended TOPSIS method confirm that Design 9 is the best choice, and after that Design 6, meaning that the conical peg with higher slope and 60% porosity is ranked better than the cylindrical peg with the same amount of porosity.

7.7 Materials and design selection of knee prosthesis—integration of QFD and MADM

This example, again, focuses on the reduction of aseptic loosening of femoral component in knee prosthesis through the design parameters including interface geometry and materials used (Jahan and Bahraminasab, 2015). At present, various designs of knee prosthesis are available with the femoral components having a similar basic design but different angles and lengths on the inner contour. These differences in the inner shape of femoral component may influence either the component or bone function adversely. To evaluate the effect of these variables, in this example design and quality tools are used with the following steps:

- identifying the design variables and planning computational experiments design,
- translating the voice of customer to technical terms and performance outputs, and indicating their relative importance using QFD,
- predicting the performance outputs of the design options made by DoE, using FEA
- finding the significant factors and evaluating the main and interaction effects of design variables through ANOVA; and then ranking the candidate design options by applying multi-attribute decision-making approach.

The steps mentioned above are presented in Fig. 7.15.

The aim is to investigate the effect of inner surface cuts (anterior, posterior and oblique surface cuts) on the biomechanical performance, whereas the outer contours of the femoral component are kept as the original design, because the complex curves of articulating surfaces are design-specific, which provide precise transitions

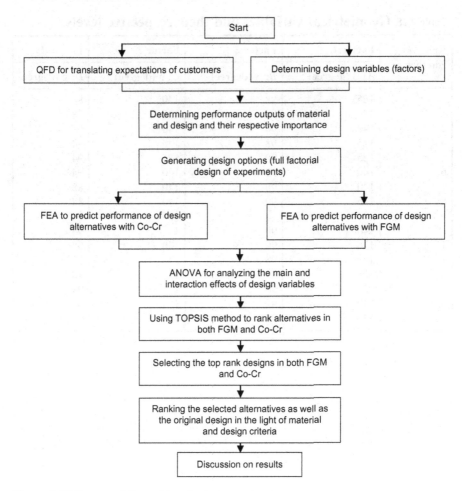

Figure 7.15 Steps used for ranking the design options.

to identify the contact characteristics with the polyethylene component. Therefore, four design variables are considered including the angle between the distal and oblique anterior surface cuts (β), the angle between the distal and anterior surface cuts (γ), the angle between the distal and oblique posterior surface cuts (δ), and the angle between the distal and posterior surface cuts (ε). These variables are altered between the ranges indicated in Table 7.28. In this example, the peg position, and the locations of points 1–4 are fixed, which leads to constant total length (L_T), length of distal cut (L_D), total height of anterior and anterior oblique surface cuts (H_A), and total height of posterior and posterior oblique surface cuts (H_P). Fig. 7.16 shows a two-dimensional (2D) outline of the femoral component, and four design variables and the constraints.

A full factorial design is applied to plan the computational runs. Since the number of factors is only four with two levels, full factorial design seems to be

Table 7.28 Geometrical variables and their respective levels

Run Number	Factor 1	Factor 2	Factor 3	Factor 4
	A: β (degree)	B: γ (degree)	C: δ (degree)	D: ε (degree)
1	125	98	140	88
2	125	98	125	88
3	140	90	140	94
4	125	98	140	94
5	125	90	125	94
6	140	98	140	94
7	140	98	140	88
8	125	90	140	88
9	125	98	125	94
10	140	90	125	94
11	140	90	140	88
12	140	98	125	94
13	125	90	125	88
14	125	90	140	94
15	140	90	125	88
16	140	98	125	88

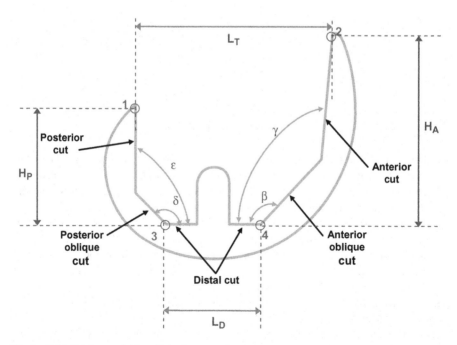

Figure 7.16 Design variables of the inner surface contour.

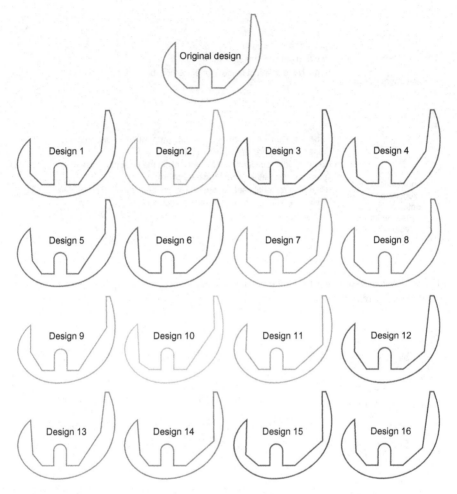

Figure 7.17 Two-dimensional profiles of generated designs.

useful, which results in 16 computational runs for each material. The geometrical variables (factors) and their levels used are shown in Table 7.28. Also, Fig. 7.17 demonstrates 2D profiles of the generated designs. Two performance outputs are considered, which are the minimum of mean stresses and the maximum of stresses STDV at several regions of interest in the bone (indicative of stress shielding). The other performance output is the maximum tangential displacement of femoral component relative to the bone (indicative of micromotion and implant instability).

For every experiment, FEA is conducted and the performance outputs are extracted. The next step is applying QFD. The expectations of customers (patients), and medical and design teams from the femoral component of knee prosthesis are

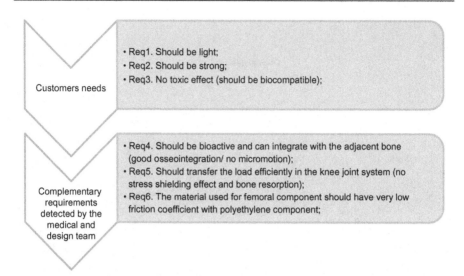

Figure 7.18 Customers' performance needs in design of femoral component and complementary requirements detected by the medical and design team.

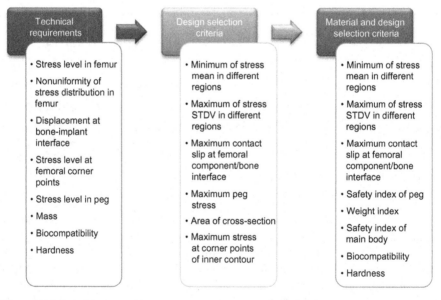

Figure 7.19 Individual matching of technical requirements with selection criteria.

presented in Fig. 7.18. Also, Fig. 7.19 illustrates how the technical requirements are categorized as design selection criteria (for ranking of design alternatives for each material), and criteria of materials and design selection (ranking of designs for both materials).

Since the aim is to maximize the minimum observed mean stress in the regions of interest, the noncompensatory model of "Maximin" is applied to consider the mean stress of the most susceptible region (stress shielding is more probable). For criterion of stresses STDV, maximum observed standard deviation in the regions is to be minimized. Other design selection criteria are also considered including maximum contact slip at the interface of femoral component and bone, area of cross-section as an index for mass (because of 2D FEA), while for the final selection, multiplication of cross-section area and density of material is considered. In the meantime, the maximum peg stress and maximum stress at corner points of inner contour are added to the criteria. Maximum peg stress is important due to the probability of peg root fracture. Maximum stress at corner points of inner contour is of importance from the stress concentration perspective, particularly for the FGM femoral component with the porous layer at the bone interface. For final selection, these two criteria are replaced by safety index of peg and safety index of main body, which are defined as the difference between yield stress of material and maximum observed stress. This is because of preventing the misleading results usually seen in traditional ratio-based performance indices (described previously in chapter: Screening of materials). Furthermore, biocompatibility and hardness are also taken into account for the final selection. Given that there is no significant subjective correlation between engineering characteristics, the roof of the house of quality was detached. The house of quality, which is used to translate the customer needs into engineering characteristics and to weighting the criteria (absolute weightings or importance), is demonstrated in Fig. 7.20. For example, for criterion of "Stress level in femur," the absolute weighting is equal to 36, which is multiplication of importance degree (4) for the related customer requirement (Req 5 in Fig. 7.18: Should transfer the load efficiently in the knee joint system) and type of relationship between customer and technical requirement (strong: 9). The last two rows in Fig. 7.20, represent the relative importance of criteria (weighting), for "design selection" (Tables 7.29 and 7.30) and "material and design selection" (Table 7.31), respectively, which is used for ranking through applying extended TOPSIS method.

Tables 7.29 and 7.30 show the design performance considered for Co-Cr alloy and FGM, respectively. Only criterion of "Minimum of mean stress in different regions" is to be maximized, and for the other criteria, less is better. In both tables, the relative weightings are obtained from the QFD absolute weighting, with regard to the criteria considered.

As can be seen from Tables 7.29 and 7.30, the Designs 3, 11, 15, and 5 are the four top ranked designs for femoral component with Co-Cr alloy, and Designs 6, 7, 3, and 1 are the top ranked designs for femoral component with FGM. These eight high ranked designs along with the original design with Co-Cr alloy are used to select the best material and design (see Table 7.30). The criteria used for selecting the best material and design are "Minimum of stress mean in different regions," "Safety index of peg," "Safety index of main body," "Biocompatibility of material," and "Hardness of interface material with PE insert," which should be maximized. For other criteria, including "Maximum of stress STDV in different regions,"

		Strong	●	+9
		Moderate	◉	+7
		Weak	○	+5

Prioritized customer requirements (What)	Degree of importance	Engineering characteristics (How)							
		Stress level in femur (A) ◁	Nonuniformity of stress distribution in femur (B) ◁	Displacement at bone-implant interface (C) ◁	Stress level at femoral corner points (F) ◁	Stress level in peg (D) ◁	Mass (E) ◁	Biocompatibility (G) ◁	Hardness (H) ◁
Req1	2						●		
Req2	3						○		
Req3	5					●		●	
Req4	4		◉		◉				
Req5	4	●		●					●
Req6	4								
Absolute weighting		36	28	36	21	27	43	45	36
Relative weighting for design selection criteria (input for Tables 7.28 and 7.29)		0.19	0.15	0.19	0.11	0.14	0.23	-	-
Relative weighting for material and design selection criteria (input for Table 7.30)		0.13	0.10	0.13	0.08	0.10	0.16	0.17	0.13

Prioritized technical descriptors

Figure 7.20 House of quality including requirements, engineering characteristics, and obtained weightings.

Table 7.29 Performance of Co-Cr alloy designs in different criteria and ranking orders scenarios by TOPSIS method

Objectives	Max.	Min.	Min.	Min.	Min.	Min.	Relative closeness to the ideal solution	Ranking
Relative weightings of design criteria	0.19	0.15	0.19	0.11	0.14	0.23		
Design options	A	B	C	D	E	F		
Design 1	2.919	3.116	34.70	24.67	900.79	115.4	0.4080	12
Design 2	3.605	3.428	44.60	24.74	918.17	114.7	0.4075	13
Design 3	1.570	2.189	20.13	26.22	804.59	72.9	0.6180	1
Design 4	1.771	2.221	21.35	26.26	925.80	108.2	0.4315	10
Design 5	2.962	2.682	35.46	27.05	901.99	73.9	0.5865	4
Design 6	4.244	1.884	45.55	27.35	874.13	109.8	0.5212	6
Design 7	1.600	1.306	20.27	27.79	849.11	122.3	0.4487	9
Design 8	1.804	1.336	21.50	27.86	867.89	86.9	0.5750	5
Design 9	1.213	2.464	87.99	17.00	934.89	101.5	0.2863	16
Design 10	1.352	1.815	97.54	25.52	813.68	79.0	0.4637	8
Design 11	0.845	0.958	61.94	10.05	779.58	69.0	0.6107	2
Design 12	0.851	1.309	62.35	10.12	883.21	109.8	0.4086	11
Design 13	1.293	2.470	90.74	17.13	885.27	85.9	0.3960	15
Design 14	1.408	1.813	99.12	25.55	892.91	68.3	0.4690	7
Design 15	0.892	1.148	63.25	10.11	796.96	68.3	0.6033	3
Design 16	0.871	1.311	60.43	10.10	866.49	116.5	0.4000	14

A, minimum of stress mean in the regions of interest (MPa); B, maximum of stress STDV in different regions (MPa); C, maximum contact slip at femoral component/bone interface (μm); D, maximum peg stress (MPa); E, area of cross-section (mm²); F, maximum stress at corner points of inner contour (MPa).

Table 7.30 **Performance of FGM designs in different criteria and ranking orders scenarios by TOPSIS method**

Objectives	Max.	Min.	Min.	Min.	Min.	Min.	Relative closeness to the ideal solution	Ranking
Relative weightings of design criteria	0.19	0.15	0.19	0.11	0.14	0.23		
Design options	A	B	C	D	E	F		
Design 1	3.100	2.655	30.61	20.16	900.79	14.4	0.6029	4
Design 2	3.818	3.449	39.58	23.54	918.17	14.3	0.5540	8
Design 3	1.707	2.208	17.93	14.09	804.59	13.1	0.6566	3
Design 4	1.925	2.245	19.16	14.34	925.8	12.2	0.6011	5
Design 5	3.150	2.668	31.24	21.26	901.99	29.0	0.4385	14
Design 6	4.466	1.866	40.36	24.17	874.13	13.8	0.6730	1
Design 7	1.741	1.249	18.03	14.71	849.11	14.7	0.6581	2
Design 8	1.961	1.349	19.25	14.97	867.89	33	0.4678	11
Design 9	1.327	2.456	79.44	12.76	934.89	12.8	0.4582	12
Design 10	1.480	1.814	87.88	15.83	813.68	18.6	0.4582	13
Design 11	0.932	1.017	57.53	7.55	779.58	22.2	0.5300	9
Design 12	0.941	1.326	57.93	7.86	883.21	12.0	0.5603	6
Design 13	1.353	2.460	83.24	12.88	885.27	17.5	0.4180	15
Design 14	1.462	1.835	90.14	15.81	892.91	18.4	0.4120	16
Design 15	0.940	1.206	58.59	7.59	796.96	22.6	0.5103	10
Design 16	0.934	1.328	58.56	7.88	866.49	13.6	0.5552	7

A, minimum of stress mean in the regions of interest (MPa); B, maximum of stress STDV in different regions (MPa); C, maximum contact slip at femoral component/bone interface (µm); D, maximum peg stress (MPa); E, area of cross-section (mm²); F, maximum stress at corner points of inner contour (MPa).

Table 7.31 **Ranking of top materials and design scenarios with regards to all technical criteria obtained from QFD**

Objectives		Max.	Min.	Min.	Max.	Min.	Max.	Max.	Max.	Relative closeness to the ideal solution	Ranking
Relative weightings		0.13	0.10	0.13	0.08	0.10	0.16	0.17	0.13		
	Design options	A	B	C	D	E	F	G	H		
Co-Cr alloy	Original	1.86	1.69	21.93	421.4	7.04	339.5	1	1	0.4531	6
	Design 3	1.570	2.189	20.13	423.78	6.68	377.1	1	1	0.4554	5
	Design 11	0.845	0.958	61.94	439.95	6.47	381	1	1	0.4142	8
	Design 15	0.892	1.148	63.25	439.89	6.61	381.7	1	1	0.4064	9
	Design 5	2.962	2.682	35.46	422.95	7.49	376.1	1	1	0.4367	7
FGM	Design 6	4.466	1.866	40.36	24.83	3.03	18.2	3	3	0.5868	1
	Design 7	1.741	1.249	18.03	34.29	2.95	17.3	3	3	0.5809	2
	Design 3	1.707	2.208	17.93	34.91	2.79	18.9	3	3	0.5584	3
	Design 1	3.100	2.655	30.61	28.84	3.13	17.6	3	3	0.5537	4

A, minimum of stress mean in the regions of interest (MPa); B, maximum of stress STDV in different regions (MPa); C, maximum contact slip at femoral component/bone interface (μm); D, safety index of peg (difference between yield of material and maximum peg stress which is 450 and 49 MPa for Co-Cr alloy and FGM, respectively); E, weight index (Area of cross-section*Density (g/cm)); F, safety index of main body (difference between yield of material and maximum stress at corner points of inner contour in which yields of 450 and 32 MPa for Co-Cr alloy and FGM were considered, respectively); G, biocompatibility of material; H, hardness of interface material with PE insert.

"Maximum contact slip at femoral component/bone interface," and "Weight index," less is better. Table 7.31 shows that Design 6 with FGM is the top rank design. From the ranking orders obtained it is indicated that the designs with FGM are better than Co-Cr alloy. The only common selected design option in the list of four top rank designs of each material is Design 3, which signifies the high contribution of selected material in the femoral component function. The interesting point is that the original design is ranked second in the designs made from Co-Cr alloy. The reason is the advantages of Co-Cr alloy (Design 3) over the original design in three criteria of "Maximum contact slip at femoral component/bone interface," "Safety index of peg," "Weight index," and "Safety index of main body." However, it appears that when considering only one single performance output (stress mean in different regions), the original design is better than Co-Cr alloy (Design 3), and it is obvious that in this condition, Design 5 is better than the original design.

7.8   Materials selection for aircraft structure repair

Aircraft can accidentally get damaged as a normal consequence of their operation, either when on the ground, for example, inadvertent heavy contact with service equipment, or when in the air, for example, a bird-strike. In order to restore the correct functionality and strength of the damaged structure, an in-service repair is often carried out. Large areas of damage will require the aircraft to be taken out of service to replace the affected part of the structure, but even if minor, damage can lead to subsequent cracking in metallic aircraft structures if left untreated. Also, when metallic aircraft are reaching the end of their service life, fatigue cracks are often found to develop along rivet holes, and other highly stressed regions of the airframe or structure such as around fasteners. In order to extend the life of aircraft, repairs can be made to arrest the cracks. Three critical steps in implementing a repair are design, choice of materials, and application. Composite doublers or repair patches provide an innovative repair technique which can enhance the way in which aircraft are maintained. Bonded repair of metallic aircraft structure is used to extend the life of flawed or underdesigned components at reasonable cost. Such repairs generally have one of three objectives: fatigue life enhancement, crack patching, and corrosion repair. The repair of a cracked structure may be performed by bonding an external patch to the structure, to either stop or slow crack growth.

The material selected for the patch must be able to withstand the expected environmental conditions in the damaged area. The patch material will almost always be either metallic or composite and within these classes are many different materials with advantages and disadvantages associated with their use (Baker et al., 2002). The processes by which the adhesive and patch materials are installed on the aircraft have a direct influence on the final properties and long-term durability of the repair. Adhesive bonding technology (Fig. 7.21), particularly bonded composite repairs, has been successfully applied by several nations to extend the lives

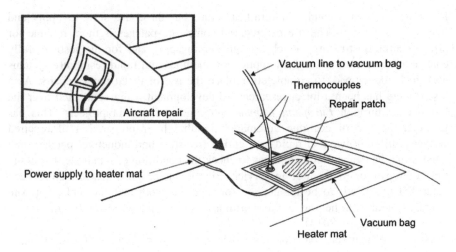

Figure 7.21 Composite bonded repair patch being applied to an aircraft structure.

of aircraft by bridging cracks in metallic structure, reducing strain levels, and repairing areas thinned by corrosion. Bonded composite reinforcements are highly efficient and cost effective when compared to conventional mechanically fastened approaches. In some cases, bonded repair technology is the only alternative to retiring a component. The repairs can be broadly divided into nonpatch procedures for minor damage and patch (or reinforcement) procedures to restore structural capability. The technique of repairing cracked metallic aircraft structures using high strength advanced composite materials is commonly known as "Crack Patching" and was pioneered by the *Aeronautical and Maritime Research Laboratories* (AMRL), for the *Royal Australian Air Force* (RAAF). The composite reinforcement, also known as the patch, can be attached to a damaged or weakened structure either by a mechanical fastener or adhesive bonding. The use of adhesively bonded composite patches as a method of repair has several advantages over mechanically fastened repair methods, which include reduced installation cost, increased strength and fatigue life and hence effective crack retardation, reduced repair down time, elimination of unnecessary fastener holes in an already weakened structure and stress concentrations at fasteners, corrosion resistance, high stiffness, and lightweight.

There is a growing trend by large companies such as *Boeing* in the United States and *Airbus* in Europe to design and manufacture modern civil aircraft with composite material structures instead of metallic material structures. This necessitates patch repair regimes that are more conducive to composite materials. Also, with the ever increasing economic pressures on civil aircraft operators to be flying as much as possible to remain competitive, the less time aircraft spend on the ground the better. This requires any repairs to aircraft to be carried out easily and quickly, including the patching of minor damage, the either permanently or temporary in the field, the latter until a permanent repair can be carried out at a scheduled service interval. Manual wet lay-up composite patches have been used for some time on secondary aircraft structure but these are slow to apply and for the resin to cure. The use of precured composite patches is more reliable for primary aircraft structure, which are quicker to apply and for the resin to fully cure. However, more automated application methods and rapid resin cure systems need to be developed, to completely replace the use of metal patch repairs. As a result, there has been a lot of research and development being conducted over the last decade through *European Commission* funded consortia projects. This has included the use of accurate automated portable laser preparation of repaired surfaces, different ways of rapid heat curing (resistive and inductive) patches, and nondestructive evaluation techniques to check the condition of repairs, either at the time of repair or to monitor in service. Significant progress has therefore been made but this needs to be maintained for repairing both metallic and composite aircraft structures to help assure airworthiness and extend economic life of aircraft in the future (Marsh, 2014).

Any patch material selection criteria will need to include: stiffness, strength, thickness, weight, conformability, operating temperature range, and product form (ie, local geometry of structure in the area of damage). The repair materials used may be conventional metals, fiber metal laminates, or composites. Metal patches are generally quicker to apply than composite patches. Thinner patches can be designed with higher modulus materials. Composite materials have higher stiffness-to-weight ratios than metals. An ability to inspect the damage through the patch is useful to monitor the integrity of repairs. Metal and fiber metal laminate patches have thermal expansion coefficients more compatible with the metal structure being repaired and are therefore more capable of enduring elevated temperatures (Duong et al., 2007). The selection criteria for patch materials will be different for repairing metallic and composite structures, with the possibility of metal or composite patch repair solutions being used on both types of structure, for example, a temporary metal patch might be used on a composite structure, which will be changed later to a permanent composite patch at the normal time of service for the aircraft. As you can see, there are advantages and disadvantages of using different patch materials but it is useful to have several options available to be able to solve the diversity of patch repair problems that might occur in practice.

The application of existing materials selection methods is unfortunately not popular in the aerospace industry (Fayazbakhsh and Abedian, 2009). This is in spite of processes being used that involve a wide range of influencing factors. As an example, consider the process of optimal materials selection for composite patch repairs in the structure of ageing metallic aircraft. The objective here is to rank the candidate materials that can be used in patch repairs (Table 7.32). The material properties considered for the design of patch repair should take into account the effects of these processes such as the cure cycle (time/temperature profile) and pressure application method used for bonding. The need is for high strength and stiffness, fatigue and environmental durability, and formability. Composite materials appear to satisfy most of these requirements.

Table 7.33 shows the required values for normalizing of data (Table 7.34). Table 7.35 shows other calculations and ranking of candidate materials. According

Table 7.32 Candidate materials, weightings, and objectives for patch repair (Fouladi et al., 2010; Baker et al., 2002; Duong and Wang, 2007; Wu and Yang, 2005)

Objectives	Max.	Min.	Max.	Max.	Max.	Min.	Max.
Subjective weightings	0.179	0.071	0.179	0.214	0.131	0.095	0.131
No. / Materials	C1	C2	C3	C4	C5	C6	C7
	Thermal expansion coefficient ($°C \times 10^{-5}$)	Approximate relative material cost	Young's modulus (GPa)	Shear modulus (GPa)	Elongation (%)	Density (g/cm³)	Tensile strength (GPa)
1 Al 2024-T3	2.32	1	73	27	18	2.78	0.483
2 Al 7075-T6	2.3	1	72	27	11	2.81	0.572
3 Titanium alloy 6Al/4V	0.9	12	110	41	14	4.5	0.95
4 Aramid/epoxy	−0.8	2	82.7	2.07	2.5	1.38	2.9
5 Glass/epoxy	0.61	1	72.5	3.52	4.8	2.58	4.03
6 HM carbon/epoxy	−1	18	390	7.1	1.8	1.81	6.9
7 Boron/epoxy (unidirectional)	0.45−2.3	40	20−208	7	0.8	2	3.4
8 Graphite/epoxy (unidirectional)	0.3−2.8	13	12−148	5	1.3	1.6	2.17
9 Aluminum laminate ARALL	1.6	7	68	17	0.89	2.3	1.282
10 Aluminum laminate GLARE	2	5	65	14.72	0.52	2.5	0.717
11 Electroformed Nickel	1.31	2	207	76	30	8.88	0.317

Table 7.33 Required parameter for normalization of patch repair materials

	C1	C2	C3	C4	C5	C6	C7
T_j	2.8	1	390	76	30	1.38	6.9
$x_j^{U_{max}}$	2.8	40	390	76	30	8.88	6.9
$x_j^{L_{min}}$	−1	1	12	2.07	0.52	1.38	0.317

Table 7.34 Normalized values of patch repair materials

No.	$V_{i,C1}^{L}$	$V_{i,C1}^{U}$	$V_{i,C2}^{L}$	$V_{i,C2}^{U}$	$V_{i,C3}^{L}$	$V_{i,C3}^{U}$	$V_{i,C4}^{L}$	$V_{i,C4}^{U}$	$V_{i,C5}^{L}$	$V_{i,C5}^{U}$	$V_{i,C6}^{L}$	$V_{i,C6}^{U}$	$V_{i,C7}^{L}$	$V_{i,C7}^{U}$
1	0.1263	0.1263	0	0	0.8386	0.8386	0.6628	0.6628	0.4071	0.4071	0.1867	0.1867	0.9748	0.9748
2	0.1316	0.1316	0	0	0.8413	0.8413	0.6628	0.6628	0.6445	0.6445	0.1907	0.1907	0.9613	0.9613
3	0.5	0.5	0.2821	0.2821	0.7407	0.7407	0.4734	0.4734	0.5427	0.5427	0.416	0.416	0.9038	0.9038
4	0.9474	0.9474	0.0256	0.0256	0.813	0.813	1	1	0.9328	0.9328	0	0	0.6076	0.6076
5	0.5763	0.5763	0	0	0.8399	0.8399	0.9804	0.9804	0.8548	0.8548	0.16	0.16	0.436	0.436
6	1	1	0.4359	0.4359	0.9788	0	0.932	0.932	0.9566	0.9566	0.0573	0.0573	0	0
7	0.6184	0.1316	1	1	1	0.4815	0.9333	0.9333	0.9905	0.9905	0.0827	0.0827	0.5317	0.5317
8	0.8158	0	0.3077	0.3077	0.8519	0.6402	0.9604	0.9604	0.9735	0.9735	0.0293	0.0293	0.7185	0.7185
9	0.3158	0.3158	0.1538	0.1538	0.8598	0.8519	0.7981	0.7981	0.9874	0.9874	0.1227	0.1227	0.8534	0.8534
10	0.2105	0.2105	0.1026	0.1026	0.8598	0.8598	0.8289	0.8289	1	1	0.1493	0.1493	0.9392	0.9392
11	0.3921	0.3921	0.0256	0.0256	0.4841	0.4841	0	0	0	0	1	1	1	1

Table 7.35 Interval numbers for S, R, and Q for patch repair materials

Materials	S-l	S-u	R-l	R-u	Q-l	Q-u	Rank
Al 2024-T3	0.513	0.513	0.150	0.150	0.279	0.279	3
Al 7075-T6	0.544	0.544	0.151	0.151	0.322	0.322	4
Titanium alloy 6Al/4V	0.572	0.572	0.133	0.133	0.249	0.249	2
Aramid/epoxy	0.733	0.733	0.214	0.214	0.944	0.944	11
Glass/epoxy	0.648	0.648	0.210	0.210	0.810	0.810	9
HM Carbon/epoxy	0.540	0.540	0.199	0.199	0.611	0.611	6
Boron/epoxy unidirectional	0.588	0.764	0.200	0.200	0.673	0.898	8
Graphite/epoxy unidirectional	0.566	0.777	0.206	0.206	0.681	0.949	10
Aluminum laminate ARALL	0.644	0.644	0.171	0.171	0.570	0.570	5
Aluminum laminate GLARE	0.644	0.644	0.177	0.177	0.611	0.611	7
Electroformed Nickel	0.385	0.385	0.131	0.131	0.000	0.000	1

to this table, metallic repairs seem to be better choices to make because their properties better match the metallic structure materials and hence simplify the design of patch. However, there are also very good reasons to consider the use of composite materials. Composites make exceptionally good repair materials due to their resistance to fatigue stresses and corrosion. The main disadvantage of using composites as patching materials however is their relatively low coefficient of thermal expansion compared to metallic structural materials. This results in undesirable residual tensile mean stresses in the area of the repair, possibly leading to distortion or even premature failure of the joint interface. This is not so much of a problem when repairing composite aircraft structures with composite patches. Of all the fiber/matrix composite combinations available, boron/epoxy, carbon/epoxy, and graphite/epoxy fiber composites are the three main options generally considered and used for patch repair materials in Australia, United Kingdom, and the United States (Fouladi et al., 2010).

The analysis described above ignores the patch joint (interface between the patch and aircraft structure), which should also be carefully designed to gradually transfer the loads (mechanical and thermal) between the patch and the structure. This can be very complicated, especially for composite structures, with multiple plies and possibly a sandwich core to deal with, depending on the depth of damage. The analysis of joint interface design is important but beyond the scope of this materials selection example.

7.9   Materials selection strategy for multiple components in an engineering product

Changing the material of one component may entail several changes in the neighboring components. For example, replacing steel with aluminum may cause galvanic corrosion with the neighboring steel components (Farag, 2002). A component-by-component materials selection procedure might lead to a suboptimal solution, therefore, selecting the optimal material mix for the different components of a product is not trivial (Leite et al., 2015). However, for simple applications, with a small number of components, traditional materials selection techniques can be employed to aid the decisions (Leite et al., 2015). This section describes the concept of multiple component materials selection using a mechanical turnbuckle as an example. In addition, the problems of economic uncertainty in materials selection problems are also highlighted.

A turnbuckle is a loop of metal with opposite internal threads at each end, to connect to the threaded ends of two ringbolts, forming a coupling that can be turned to tighten or loosen the tension in members attached to the ringbolt (Farag, 2008). Fig. 7.22 shows a typical form of turnbuckle. Turnbuckles are most commonly used in applications with cables/wires under tension. Examples includes aircraft control cables, yacht rigging, professional wrestling/boxing rings, and piping/ducting.

The factors that affect the performance in service include:

- the yield strength (YS) and ultimate tensile strength (UTS) of the loop or one of the ringbolts
- the screw threads shearing, or stripping on the loop or on one of the ringbolts
- fatigue fracture of the loop or one of the ringbolts
- creep strain in the loop or one of the ringbolts
- fracture of the loop or one of the ring bolts due to excessive loading of the system or as a result of impact loading
- corrosion as a result of environmental attack and galvanic action between ringbolt and loop when they are made of different metals.

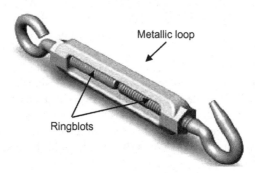

Figure 7.22 Turnbuckle showing main components: ringbolts and loop.

Table 7.36 **Hypothetical candidate materials and related material performance dependent criteria for ringbolt**

Weighting	0.2	0.2	0.2	0.2	0.2
Material pair	UTS (MPa)	YS (MPa)	FS (MPa)	Corrosion resistance	Mass (g)
M_{R1}-M_{L1}	135	90	80	2	300
M_{R1}-M_{L2}	350	130	145	1	400
M_{R1}-M_{L3}	240	75	75	3	500
M_{R2}-M_{L1}	135	90	80	2	750
M_{R2}-M_{L2}	460	185	170	2	850
M_{R2}-M_{L3}	240	75	75	4	950

It is assumed that the turnbuckle will not be subjected to high-temperature service conditions therefore creep strain failure will not be a concern. Copper-base alloys, Aluminum alloys, and Steels can be candidate materials for the ringbolt. Gray cast irons (ASTM A48-74), Nodular cast irons (ASTM A536), Aluminum alloys, and Copper-base alloys can be candidate materials for the loop. For the present case study, it is presumed that there are two candidate materials for the ringbolt and three candidate materials for the loop. For each loop and ringbolt material pair, hypothetical data is generated randomly and shown in Table 7.36. It is wished to determine optimum combination of materials. Both tensile and fatigue loads act on the turnbuckle. Due to these loads, the turnbuckle may fail if the loop or one of the ringbolts fails. Therefore, when different materials are used for the ringbolt and the loop, the lower UTS, fatigue strength (FS), and YS are used to rate the entire system. Each pair of candidate materials is assigned a corrosion resistance rating, where 1 = poor, 2 = fair, 3 = good, and 4 = very good.

Cost is uncertain in many aspects and it is sometimes essential to define "capital cost" and "variable cost" as separate and distinct attributes (Thurston and Locascio, 1994). Capital or fixed costs are costs that are independent of the output. Variable costs are costs that vary with the output. Generally, variable and fixed costs have an opposite relationship. For example, steel typically has significantly larger capital cost requirements than composites, but lower variable cost. Cost estimates are decisions about what costs will be in the future. The accuracy of the estimate is inversely proportional to the span of time between the estimate and the event. Similar all decisions, cost estimates can be divided to those made under certainty, risk, or uncertainty (Asiedu and Gu, 1998). Uncertainties are related to the estimation process in terms of data, estimation techniques, and scenarios analysis.

In particular, cost estimation of design for the conceptual and embodiment stages of the design process are not supported sufficiently. Fig. 7.23 presents the structure

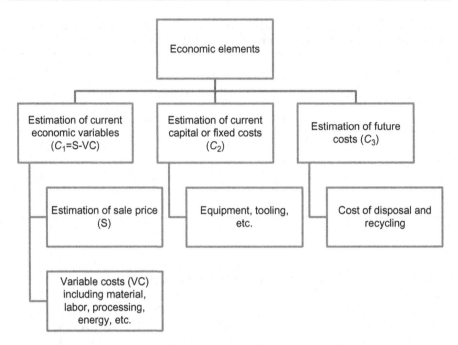

Figure 7.23 Economic elements that can be estimated by designers in materials selection stage.

of economic elements that can be related to the selection of material. Precalculation in economic analysis is useful for the early stages of design and aims to make effective choices between material alternatives based on a view of the different costs associated with each alternative.

Table 7.37 shows the estimated costs for hypothetical loop and ringbolt material pair in the form of relative cost. It is assumed that life duration of a material is related to technical performance (eg, fatigue durability). Relative costs are based on the minimum value of each column. The ability to estimate the costs with known uncertainty will provide grounds for making better decisions, and therefore, offer distinct engineering and business advantages (Goh et al., 2010). Therefore, the data presented in the form of interval. For each pair of materials, cost of equipment and tooling, disposal and recycling, and variable cost, including material, labor, and energy, are considered. The importance of each type of cost is presented in Table 7.37. Capital and variable costs are defined as separate attributes. This enables the designer to fine-tune the trade-off analysis.

Fig. 7.24 shows the overall steps for improving the process of multiple materials selection using interval target-based VIKOR (described previously in chapter: Multi-attribute decision-making for ranking of candidate materials), and with specific attention to economic criteria.

The economic index may conflict with the performance index; hence, some trade-off between the criteria is needed to ensure satisfactory materials selection.

Table 7.37 Hypothetical candidate materials and related economic selection criteria for ringbolt

	Estimation of relative costs		
Weighting	**0.25**	**0.4**	**0.35**
Material pair	**Cost of disposal and recycling**	**Cost of equipment and tooling**	**Variable cost including material, labor, and energy**
M_{R1}-M_{L1}	1	1.7−2	2.5−2.7
M_{R1}-M_{L2}	1.2−1.3	1.2−2.2	2.7−2.8
M_{R1}-M_{L3}	1.3−1.5	1	3−3.3
M_{R2}-M_{L1}	2.3−2.4	2.8−3.1	1.4−1.5
M_{R2}-M_{L2}	2.5−2.7	3−4	1
M_{R2}-M_{L3}	2.7−2.9	2.5−2.7	1.2−1.3

The Pareto line or curve is defined as the border between the dominated and nondominated candidate materials and can help designers find a single solution between the different objectives. The borderline limits beyond which the design cannot be further improved. The concept of a Pareto line or curve was described previously in chapter "Multi-criteria decision-making for materials selection". If this border is shown as a function of the objectives, it is a continuous line, which contains possible optimal combinations of the objectives.

Table 7.38 demonstrates ranking of alternative and performance index of Q using VIKOR method. The lower the index value, the better the performance of the alternative. Table 7.39 shows upper and lower range economic index (Q) using interval VIKOR method, the average of Q is used as final index.

The designer might strive to design the perfect component, but the price must be competitive in the commercial marketplace, therefore the best candidate material would be a trade-off between the performance and economic indexes. Fig. 7.25 shows a Pareto front diagram. It is apparent from Fig. 7.25 that three material pairs are dominated by other alternatives. It means M_{R1}-M_{L2} dominates alternatives M_{R1}-M_{L3}, M_{R2}-M_{L3}, and M_{R2}-M_{L1}. Overall, these results indicate that the designer can select a pair of materials from Pareto points (M_{R2}-M_{L2}, M_{R1}-M_{L2}, and M_{R1}-M_{L1}).

The main goal of this example was to explain the concept of multiple component materials selection/substitution. Also, the present study made several noteworthy contributions to available methods. These include developing a formal method for taking into account an estimation of capital costs in the earlier stages of the design process. This results in business advantages by combining the economic elements using MADM rather than simple summation of current and future costs, taking into account uncertainty in economic elements using interval data, and removing dominated solutions using Pareto front diagram.

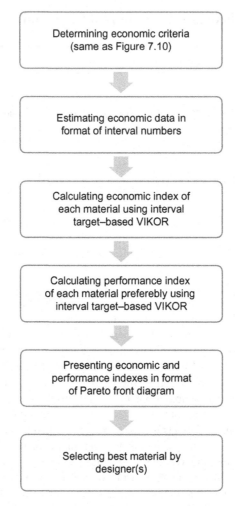

Figure 7.24 Steps used for multiple materials selection process associated with economic uncertainty.

Table 7.38 **Performance index for material pairs using VIKOR method**

Material pair	S	R	Q
M_{R1}-M_{L1}	0.696	0.200	0.870
M_{R1}-M_{L2}	0.451	0.200	0.640
M_{R1}-M_{L3}	0.664	0.200	0.840
M_{R2}-M_{L1}	0.834	0.200	1.000
M_{R2}-M_{L2}	0.303	0.169	0.000
M_{R2}-M_{L3}	0.735	0.200	0.907

Table 7.39 Economic index for material pairs using interval VIKOR method

Material pair	S-l	S-u	R-l	R-u	Q-l	Q-u	Avg of Q
M_{R1}-M_{L1}	0.392	0.403	0.259	0.263	0.115	0.146	0.130
M_{R1}-M_{L2}	0.343	0.497	0.274	0.298	0.080	0.429	0.254
M_{R1}-M_{L3}	0.394	0.416	0.350	0.350	0.423	0.461	0.442
M_{R2}-M_{L1}	0.534	0.627	0.280	0.360	0.436	0.867	0.651
M_{R2}-M_{L2}	0.621	0.624	0.400	0.400	0.988	0.994	0.991
M_{R2}-M_{L3}	0.512	0.596	0.250	0.300	0.296	0.611	0.454

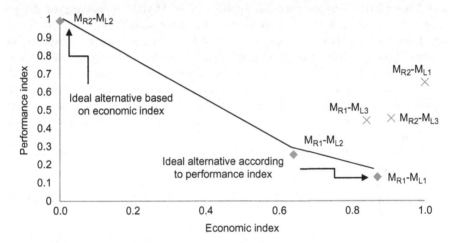

Figure 7.25 Pareto points and dominated solutions.

7.10 Material and design selection of thin-walled cylindrical tubes in the presence of data uncertainty

The global transportation industry is constantly looking for lightweight products to reduce energy consumption. In this context, several innovative structures and lightweight materials are introduced for use in an automobile body, replacing the traditional steel components (eg, aluminum alloys). In fact, thin-walled cylindrical tubes can potentially be used in all types of vehicle and moving components, including road vehicles, trains, aircraft, ships, lifts and machinery to protect both passengers and the structure itself during impact. The two important necessities for crush analysis of thin-walled cylindrical tubes are to be lightweight and possess acceptable crashworthiness. It was found that among all cross-sections, circular tubes provide the most energy absorption capacity and mean crushing load. In recent times, thin-walled structures have been filled with foams, which are a new class of materials having extremely low densities, good combination of mechanical properties, and improved energy absorption that aid in stability of the tubes. In the design of thin-walled tubes, the maximum force (highest load) required to cause a large amount of permanent deformation, is as important as the energy absorption because of the following two reasons:

- No permanent deformation is acceptable in low-speed and low-energy impacts.
- The peak load directly influences the force on the vehicle occupants (in actual physical tests the peak load is measured from the reaction force).

This means that to gain the best crashworthiness design as well as being lightweight, various criteria must be met simultaneously. This example (Rezvani and Jahan, 2015) demonstrates the application of MADM in ranking of design scenarios in the presence of data uncertainty. Thin-walled tubes with different geometry and materials have been widely used as collapsible energy absorbers in various types of structural applications. Here, the idea of applying an initiator on foam-filled thin-walled circular tubes having stiffened annular rings is used to create design alternatives, which is examined by both experiments and FEA. When a thin-walled structure is collapsed under axial compression load, the energy absorption characteristics are measured through the following parameters, which are obtained with reference to Fig. 7.26.

The total energy absorbed E_{abs}, in crushing the structure is equal to the area under the load–displacement curve, where:

$$E_{abs} = \int P d\delta \tag{7.16}$$

In the above equation, P and δ are the crush force and crush distance, respectively.

The mean crush load is defined as the ratio of absorbed energy E_{abs} and total deformation (δ):

$$P_m = \frac{E_{abs}}{\delta} \tag{7.17}$$

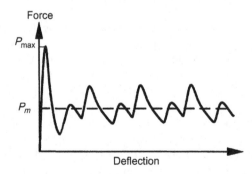

Figure 7.26 Typical load–displacement curve.

The crush force efficiency (CFE) is obtained by the equation:

$$CFE = \frac{P_m}{P_{max}} \tag{7.18}$$

Here, P_m and P_{max} denote, in respect, the mean crush load and initial peak load. The ideal energy absorber is expected to immediately achieve a maximum load and maintain it for the entire component length.

Specific energy absorption (SEA) is one of the most key characteristics of energy absorbers. This is used as a criterion when designing the structures, in which the weight efficiency is the main goal. SEA offers a way of comparison of energy absorption capacity of structures with different masses and can be calculated by Eq. (7.19):

$$SEA = \frac{E_{abs}}{m} \tag{7.19}$$

where m is the total mass of the specimen/structure undergoing deformation

The effect of initiator (applied at the top of the foam-filled aluminum tube), various densities of rigid polyurethane foam, and number of annular rings under axial compression load are demonstrated through energy absorption, crush force efficiency (CFE), and mass of structure. Using these variables, there is an expectation to see an increase in the SEA and therefore a reduction of the suddenly force applied to the main part of the automotive being experienced by its occupants. FEA is first carried out and the performance outputs (criteria) extracted. To verify the numerical results obtained, several quasistatic axial compression experiments are performed, and both load–displacement curves and the deformation mechanism of the structure are analyzed. However, there is usually a difference between the FEA predictions and real experimental measurements. Hence, after minimizing the errors in the acceptable range, designers have to rely either on simulation outputs or on experimental data results when missing some vital information in selecting optimal materials, shapes and dimensions for the design of components. It appears the extended version of interval VIKOR method (described previously in chapter: Multi-attribute decision-making for ranking of candidate materials) can be fitted well to the selection problem of lightweight design in thin-walled cylindrical tube. Fig. 7.27 demonstrates the decision-making process.

The test specimens are made of the aluminum alloy 6061-T6. Two different rigid polyurethane (PU) foams are applied in the experiments, with the densities of $\rho_1 = 174$ (kg/m^3) and $\rho_2 = 217$ (kg/m^3). Fig. 7.28 shows a schematic of the design and fixed dimensions. which are $D_i = 42$, $D_o = 50$, $L = 110$, $t = 1.5$, $d = 2.5$. The initiators are made of steel of length 20 mm (one-initiator design) and 10 mm (two-initiator design). In order to avoid any contact between the tube and the initiators, the diameter of the initiator should be less than the inner diameter of the tube. In fact, the initiators play the role of merely pressing on the foam. The dimensions and more details about the manufactured specimens and initiators are provided in Table 7.40. The terms represent the number of rings and initiators, respectively, for

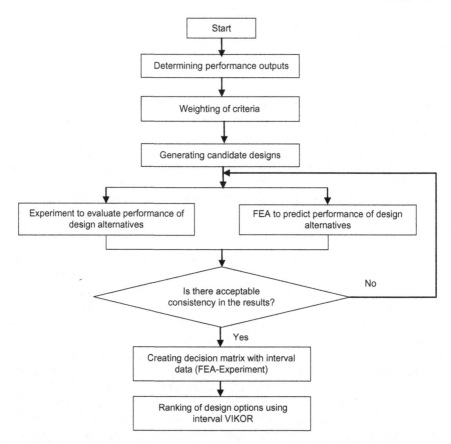

Figure 7.27 Proposed steps for design decision-making process in presence of both numerical and physical experiments.

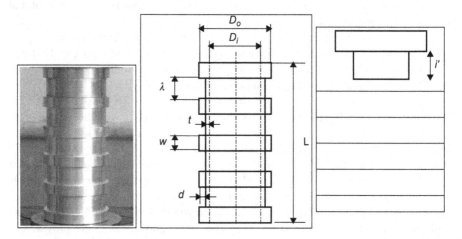

Figure 7.28 Cylinder tube stiffened by annular rings and initiator with its detailed design.

Table 7.40 **Experimental and numerical results after axial crushing**

Density (kg/m³)	Specimen	m (kg)	Simulation				Experimental				Error%	
			E (J)	P_{max} (kN)	P_m (kN)	CFE (%)	E (J)	P_{max} (kN)	P_m (kN)	CFE (%)	E (J)	CFE
217	C-4R-1I	0.428	1917.04	52.58	29.35	56	1962.80	53.31	29.74	56	2.33	0.00
	C-5R-0I	0.125	1698.15	55.28	34.94	59	1781.21	56.54	33.60	59	4.66	0.00
	C-5R-1I	0.437	2069.47	60.11	29.33	49	2128.64	59.69	29.15	49	2.78	0.00
	C-6R-1I	0.445	2029.35	51.67	26.01	50	2011.19	50.10	25.48	51	−0.90	1.96
174	C-4R-1I	0.421	1677.81	53.30	23.19	43	1768.38	53.82	23.27	43	5.12	0.00
	C-5R-1I	0.430	1818.46	54.11	25.14	46	1915.23	54.82	24.82	45	5.05	−2.22
	C-6R-1I	0.438	1581.51	42.52	22.24	52	1684.16	44.46	22.75	51	6.10	−1.96
	C-6R-2I	0.534	2635.99	59.47	36.74	62	2656.34	54.30	33.02	61	0.77	−1.64

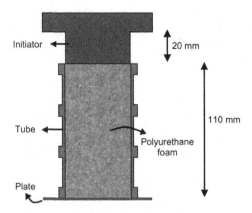

Figure 7.29 A typical arrangement of thin-walled tube between the steel initiator and plate.

example, code of "C-4R-1I" represents a cylindrical sample with 4 rings and one initiator.

As indicated in Fig. 7.29, the specimen is situated between the steel initiator and plate of the testing machine with no additional fixing. The initiator acts in two stages: (1) the initiator is moved with length *of $l' = 20$ or 10 mm for two-initiator design, and only presses the foam, (2) when the initiator reaches to the top of the tube, the tube's buckling is initiated.

The results from experimental tests and numerical simulations for energy absorption, maximum crushing load, mean crushing load, and crush force efficiency are shown in Table 7.40. It should be noted that in Table 7.40 the total weight of the structure including the mass of the tube with foam and initiator is considered. Also, it can be seen that there is an acceptable agreement between FEA and experimental tests.

The data in Table 7.40 show that the C-6R-1I specimen with density of 174 kg/m³ has lower energy absorption than the other specimens. Fig. 7.30 presents the load−displacement curves from the experimental test for C-6R-2I and C-6R-1I specimens. These specimens are filled with PU foam with density of 174 kg/m³. C-6R-2I has 6 rings and two initiators with length of 10 mm (at the top and bottom of the tube). Higher energy absorption is seen for the two-initiators-specimen compared to the C-6R-1I specimen. As can be found from Table 7.40, not only the initial peak load in C-6R-2I specimen increased up to 18% compared to the specimen of C-6R-1I, but also the SEA and the crush force efficiency increased up to 36% and 16%, respectively. Thus, it can be concluded that the C-6R-2I specimen reveals more appropriate results in term of energy absorption.

Table 7.41 represents the design alternatives and objectives, selection criteria, including mass, energy absorption, and CFE. Equal weighting is assigned to the considered criteria. Data for criteria of energy absorption, and CFE are considered interval demonstrating the range between experimental and simulation values.

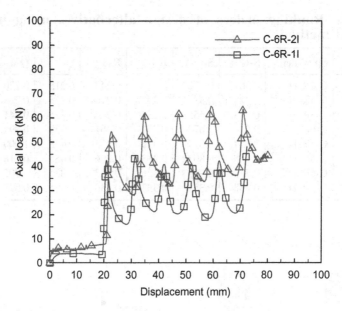

Figure 7.30 Comparison between load—displacement curves obtained from experimental test for foam-filled specimens with density of $\rho = 174$ kg/m^3.

Table 7.41 **Design selection criteria and their importance**

Objective		Minimum	Maximum	Maximum
Weighting		0.333	0.333	0.333
Density	Specimen	*m* (kg)	*E* (J)	*CFE* (%)
217	C-4R-1I	0.428	1917–1963	56
217	C-5R-0I	0.125	1698–1781	59
217	C-5R-1I	0.437	2069–2129	49
217	C-6R-1I	0.445	2011–2029	50–51
174	C-4R-1I	0.421	1678–1768	43
174	C-5R-1I	0.430	1818–1915	45–46
174	C-6R-1I	0.438	1582–1684	51–52
174	C-6R-2I	0.534	2636–2656	61–62

The ranking orders of design scenarios using interval VIKOR approach is shown in Table 7.42. The most prominent result to emerge from the ranking is that the 4 rings specimen with initiator and higher density (C-4R-1I) is the top rank design.

The design ranking orders indicate that the specimens with higher density of foam perform better than the low-density specimens. The interesting point is that for five rings and density of 217 kg/m^3, the initiator is not beneficial. Fig. 7.31

Table 7.42 Ranking orders of design alternatives using interval VIKOR method

Density	Specimen	S-l	S-u	R-l	R-u	Q-l	Q-u	Rank
217	C-4R-1I	0.561	0.570	0.247	0.247	0.213	0.220	1
217	C-5R-0I	0.328	0.333	0.293	0.296	0.268	0.293	2
217	C-5R-1I	0.638	0.655	0.254	0.254	0.325	0.341	3
217	C-6R-1I	0.633	0.672	0.261	0.261	0.358	0.394	4
174	C-4R-1I	0.854	0.877	0.315	0.333	0.878	1.000	7
174	C-5R-1I	0.758	0.803	0.263	0.296	0.486	0.718	6
174	C-6R-1I	0.739	0.773	0.326	0.333	0.835	0.905	8
174	C-6R-2I	0.333	0.357	0.333	0.333	0.505	0.527	5

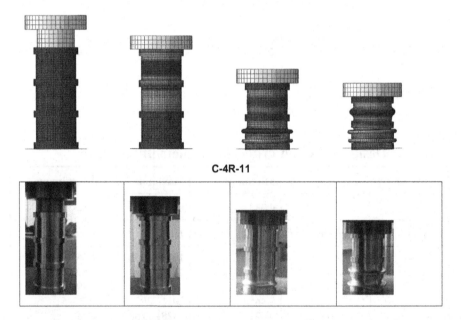

C-4R-11

Figure 7.31 Crushed shapes of specimens derived from numerical simulation and physical experiments.

shows the 217-C-4R-1I specimen after the numerical simulation and the experimental crushing test. Fig. 7.31 also shows the annular rings on the outer surface dividing the thin-walled tube into several shorter thin-walled portions. The collapsing mode of the tube, therefore, can be controlled, with the annular rings on the outer surface of the tube causing uniform plastic hinges.

Fig. 7.32 shows the comparison of SEA as a function of number of rings for two densities obtained in the experiments. It is clear from Fig. 7.32 that the SEA is raised with increasing density of PU foam from 174 to 217 kg/m^3. The SEA for

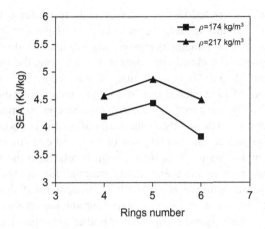

Figure 7.32 Effect of ring number on the specific energy absorption of foam-filled tubes with initiator.

four rings (C-4R-1I) and five rings (C-5R-1I) with one initiator and density of 217 kg/m^3 is 4.58 and 4.87 kJ/kg, respectively, even though for specimen C-6R-1I the SEA is 4.51 kJ/kg. It should be pointed out that by increasing the number of rings, the tube mass is increased. Meanwhile, Fig. 7.32 demonstrates that by considering the same importance for both energy absorption and mass of structure, the design with 5 rings and density of 217 kg/m^3 is the best option. Given that in this way it is not possible to also take into account the CFE concurrently, the result is different from multi-criteria analysis using the interval VIKOR approach, which suggests 217-C-4R-1I as the top rank design.

Although the two-initiator-structure design is the best design among all of the tube profiles examined when only considering the energy absorption and CFE, the mass of the structure would be too high. The interval VIKOR method suggests that the design with an initiator, four rings, and higher density of foam has superior performance in all criteria, although this finding may not be extendable to other tube geometries and materials.

7.11 Group decision-making in materials selection

This example describes the materials selection for a motorcar instrument panel (Jeya Girubha and Vinodh, 2012) using Fuzzy Group MADM. The main requirements or criteria, in this example, for the selection of instrumental panel materials includes maximum temperature limit, recyclability, elongation, weight, thermal conductivity, tensile strength, cost, and toxicity level. Most instrumental panel components are subjected to relatively high temperatures, which can cause squeaking and rattling behavior. Therefore, appropriate thermal properties are required for the material used in this application to avoid vibrations. On the other hand, the panel should be correctly connected to the motorcar body to decrease the vibration and the relative motion. In this regard, the material should possess limited elongation (minimum). A low-stiffness material may readily vibrate and subsequently relative motion occurs between the mated components. Furthermore, the thermal conductivity of the material should be low because of the occurrence of stick-slip. The mechanical strength of the material should be high enough to resist heavy load and high impacts. Also, the material is required to maintain its mechanical strength under high temperature. The weight of the material is also important because lightweight components reduce the consumption of fuel, and consequently improves the fuel economy of motorcars (40% of total weight is related to the body and interior components). The other requirement of the material for the instrument panel is recyclability, which is economically beneficial particularly if the material can be recycled back into another high purity material, producing another component with the same quality as the original component. Another important issue is the toxicity of the material, with consideration needed to be given to the potential human health effects of the materials.

The trapezoidal fuzzy numbers (described previously in chapter: Multi-attribute decision-making for ranking of candidate materials) that are used for the ratings of alternatives with respect to various criteria are shown in Table 7.43.

For this example, the terms for describing the importance of material with respect to criteria assessed by decision-makers are shown in Tables 7.44 and 7.45. Let the fuzzy rating of the kth decision-maker for material i with respect to criterion j, be $X_{ijk} = \{X_{ijk1}, X_{ijk2}, X_{ijk3}, X_{ijk4}\}$.

As shown in Table 7.46, the aggregated fuzzy ratings X_{ij} of alternatives with respect to each criterion are calculated using Eq. (7.20).

$$X_{ij} = \{X_{ij1}, X_{ij2}, X_{ij3}, X_{ij4}\} \tag{7.20}$$

where $X_{J1} = \min_k\{X_{jk1}\}$, $X_{J2} = \frac{1}{k}\sum X_{jk2}$, $X_{J3} = \frac{1}{k}\sum X_{jk3}$, $X_{J4} = \max_k\{X_{jk4}\}$

To remove the dimensions of all criteria, normalization can be used. To avoid complexity in mathematical operations in the decision process, the linear scale transformation is used to transform the five various criteria scales into comparable scales.

Table 7.43 Linguistic terms and corresponding fuzzy numbers for each material

Linguistic variable	Fuzzy number
Very low (VL)	(0.0, 0.0, 0.1, 0.2)
Low (L)	(0.1, 0.2, 0.2, 0.3)
Fairly low (FL)	(0.2, 0.3, 0.4, 0.5)
Medium (M)	(0.4, 0.5, 0.5, 0.6)
Fairly high (FH)	(0.5, 0.6, 0.7, 0.8)
High (H)	(0.7, 0.8, 0.8, 0.9)
Very high (VH)	(0.8, 0.9, 1.0, 1.0)

Table 7.44 Rating of material with respect to criteria assessed by five decision-makers (linguistic variable)

		C1	C2	C3	C4	C5	C6	C7	C8
D1	M1	FH	H	FH	FH	H	FH	FH	FH
	M2	M	FH	H	H	H	FH	H	FH
	M3	VH	H	VH	H	H	VH	H	VH
	M4	H	VH	H	H	VH	H	VH	H
D2	M1	FH	H	FH	FH	FH	H	FH	FH
	M2	FH	M	FH	M	FH	M	M	M
	M3	H	VH	H	VH	H	VH	H	H
	M4	H	H	VH	H	VH	H	VH	VH
D3	M1	FH	H	FH	H	FH	FH	H	FH
	M2	FH	FH	H	H	H	H	H	H
	M3	VH	VH	H	H	VH	VH	H	H
	M4	H	VH	H	H	H	VH	H	H
D4	M1	H	FH	H	FH	FH	H	FH	H
	M2	FH	FH	M	FH	M	FH	M	FH
	M3	H	VH	H	VH	H	H	VH	H
	M4	H	H	VH	H	H	H	VH	VH
D5	M1	FH	H	FH	H	FH	H	FH	FH
	M2	H	FH	H	FH	H	FH	M	M
	M3	VH	H	VH	H	VH	H	VH	H
	M4	H	VH	H	H	H	VH	H	M

The set of criteria can be divided into benefit criteria (the larger the rating, the greater the preference) and cost criteria (the smaller the rating, the greater the preference). Therefore, the normalized fuzzy decision matrix is represented in Table 7.47, with regards to cost criterion (C) and benefit criterion (B) and using Eqs. (7.21) and (7.22).

$$r_{ij} = \left\{ \frac{X_{ij1}}{X_{j4}^+}, \frac{X_{ij2}}{X_{j4}^+}, \frac{X_{ij3}}{X_{j4}^+}, \frac{X_{ij4}}{X_{j4}^+} \right\} \quad j \in B \tag{7.21}$$

$$r_{ij} = \left\{ \frac{X_{ij1}^-}{X_{j1}^-}, \frac{X_{ij2}}{X_{j1}^-}, \frac{X_{ij3}}{X_{j1}^-}, \frac{X_{ij4}}{X_{j1}^-} \right\} \quad j \in C \tag{7.22}$$

Where $X_{j4}^+ = \max_i \{\text{decision matrix}\}, j \in B$

$X_{j1}^- = \min_i \{\text{decision matrix}\}, j \in C$

The Crisp values of decision matrix are obtained using Eq. (5.8) and are shown in Table 7.48. Using the data in Table 7.48, the ranking orders of materials will be M3-M4-M1-M2 either by original VIKOR method or the new version that was explained previously in chapter: Multi-attribute decision-making for ranking of candidate materials.

Table 7.45 Rating of material with respect to criteria assessed by five decision-makers (fuzzy set)

		C1	C2	C3	C4	C5	C6	C7	C8
D1	M1	(0.5, 0.6, 0.7, 0.8)	(0.7, 0.8, 0.8, 0.9)	(0.5, 0.6, 0.7, 0.8)	(0.5, 0.6, 0.7, 0.8)	(0.7, 0.8, 0.8, 0.9)	(0.5, 0.6, 0.7, 0.8)	(0.5, 0.6, 0.7, 0.8)	(0.5, 0.6, 0.7, 0.8)
	M2	(0.4, 0.5, 0.5, 0.6)	(0.5, 0.6, 0.7, 0.8)	(0.7, 0.8, 0.8, 0.9)	(0.7, 0.8, 0.8, 0.9)	(0.7, 0.8, 0.8, 0.9)	(0.5, 0.6, 0.7, 0.8)	(0.7, 0.8, 0.8, 0.9)	(0.5, 0.6, 0.7, 0.8)
	M3	(0.8, 0.9, 1, 1)	(0.7, 0.8, 0.8, 0.9)	(0.8, 0.9, 1, 1)	(0.7, 0.8, 0.8, 0.9)	(0.7, 0.8, 0.8, 0.9)	(0.8, 0.9, 1, 1)	(0.7, 0.8, 0.8, 0.9)	(0.8, 0.9, 1, 1)
	M4	(0.7, 0.8, 0.8, 0.9)	(0.8, 0.9, 1, 1)	(0.7, 0.8, 0.8, 0.9)	(0.7, 0.8, 0.8, 0.9)	(0.8, 0.9, 1, 1)	(0.7, 0.8, 0.8, 0.9)	(0.8, 0.9, 1, 1)	(0.7, 0.8, 0.8, 0.9)
D2	M1	(0.5, 0.6, 0.7, 0.8)	(0.7, 0.8, 0.8, 0.9)	(0.5, 0.6, 0.7, 0.8)	(0.5, 0.6, 0.7, 0.8)	(0.5, 0.6, 0.7, 0.8)	(0.7, 0.8, 0.8, 0.9)	(0.5, 0.6, 0.7, 0.8)	(0.5, 0.6, 0.7, 0.8)
	M2	(0.5, 0.6, 0.7, 0.8)	(0.4, 0.5, 0.5, 0.6)	(0.5, 0.6, 0.7, 0.8)	(0.4, 0.5, 0.5, 0.6)	(0.5, 0.6, 0.7, 0.8)	(0.4, 0.5, 0.5, 0.6)	(0.4, 0.5, 0.5, 0.6)	(0.4, 0.5, 0.5, 0.6)
	M3	(0.7, 0.8, 0.8, 0.9)	(0.8, 0.9, 1, 1)	(0.7, 0.8, 0.8, 0.9)	(0.8, 0.9, 1, 1)	(0.7, 0.8, 0.8, 0.9)	(0.8, 0.9, 1, 1)	(0.7, 0.8, 0.8, 0.9)	(0.7, 0.8, 0.8, 0.9)
	M4	(0.7, 0.8, 0.8, 0.9)	(0.8, 0.9, 1, 1)	(0.8, 0.9, 1, 1)	(0.7, 0.8, 0.8, 0.9)	(0.8, 0.9, 1, 1)	(0.7, 0.8, 0.8, 0.9)	(0.8, 0.9, 1, 1)	(0.8, 0.9, 1, 1)
D3	M1	(0.5, 0.6, 0.7, 0.8)	(0.7, 0.8, 0.8, 0.9)	(0.5, 0.6, 0.7, 0.8)	(0.7, 0.8, 0.8, 0.9)	(0.5, 0.6, 0.7, 0.8)	(0.5, 0.6, 0.7, 0.8)	(0.7, 0.8, 0.8, 0.9)	(0.5, 0.6, 0.7, 0.8)
	M2	(0.5, 0.6, 0.7, 0.8)	(0.5, 0.6, 0.7, 0.8)	(0.7, 0.8, 0.8, 0.9)	(0.7, 0.8, 0.8, 0.9)	(0.7, 0.8, 0.8, 0.9)	(0.7, 0.8, 0.8, 0.9)	(0.7, 0.8, 0.8, 0.9)	(0.7, 0.8, 0.8, 0.9)
	M3	(0.8, 0.9, 1, 1)	(0.8, 0.9, 1, 1)	(0.7, 0.8, 0.8, 0.9)	(0.7, 0.8, 0.8, 0.9)	(0.8, 0.9, 1, 1)	(0.8, 0.9, 1, 1)	(0.7, 0.8, 0.8, 0.9)	(0.7, 0.8, 0.8, 0.9)
	M4	(0.7, 0.8, 0.8, 0.9)	(0.8, 0.9, 1, 1)	(0.7, 0.8, 0.8, 0.9)	(0.7, 0.8, 0.8, 0.9)	(0.8, 0.9, 1, 1)	(0.8, 0.9, 1, 1)	(0.7, 0.8, 0.8, 0.9)	(0.7, 0.8, 0.8, 0.9)
D4	M1	(0.7, 0.8, 0.8, 0.9)	(0.7, 0.8, 0.8, 0.9)	(0.7, 0.8, 0.8, 0.9)	(0.5, 0.6, 0.7, 0.8)	(0.5, 0.6, 0.7, 0.8)	(0.7, 0.8, 0.8, 0.9)	(0.5, 0.6, 0.7, 0.8)	(0.7, 0.8, 0.8, 0.9)
	M2	(0.5, 0.6, 0.7, 0.8)	(0.5, 0.6, 0.7, 0.8)	(0.4, 0.5, 0.5, 0.6)	(0.5, 0.6, 0.7, 0.8)	(0.4, 0.5, 0.5, 0.6)	(0.5, 0.6, 0.7, 0.8)	(0.4, 0.5, 0.5, 0.6)	(0.5, 0.6, 0.7, 0.8)
	M3	(0.7, 0.8, 0.8, 0.9)	(0.8, 0.9, 1, 1)	(0.7, 0.8, 0.8, 0.9)	(0.8, 0.9, 1, 1)	(0.7, 0.8, 0.8, 0.9)	(0.7, 0.8, 0.8, 0.9)	(0.8, 0.9, 1, 1)	(0.7, 0.8, 0.8, 0.9)
	M4	(0.7, 0.8, 0.8, 0.9)	(0.7, 0.8, 0.8, 0.9)	(0.8, 0.9, 1, 1)	(0.7, 0.8, 0.8, 0.9)	(0.7, 0.8, 0.8, 0.9)	(0.7, 0.8, 0.8, 0.9)	(0.8, 0.9, 1, 1)	(0.8, 0.9, 1, 1)
D5	M1	(0.5, 0.6, 0.7, 0.8)	(0.7, 0.8, 0.8, 0.9)	(0.5, 0.6, 0.7, 0.8)	(0.7, 0.8, 0.8, 0.9)	(0.5, 0.6, 0.7, 0.8)	(0.7, 0.8, 0.8, 0.9)	(0.5, 0.6, 0.7, 0.8)	(0.5, 0.6, 0.7, 0.8)
	M2	(0.7, 0.8, 0.8, 0.9)	(0.5, 0.6, 0.7, 0.8)	(0.7, 0.8, 0.8, 0.9)	(0.5, 0.6, 0.7, 0.8)	(0.7, 0.8, 0.8, 0.9)	(0.5, 0.6, 0.7, 0.8)	(0.4, 0.5, 0.5, 0.6)	(0.4, 0.5, 0.5, 0.6)
	M3	(0.8, 0.9, 1, 1)	(0.7, 0.8, 0.8, 0.9)	(0.8, 0.9, 1, 1)	(0.7, 0.8, 0.8, 0.9)	(0.8, 0.9, 1, 1)	(0.7, 0.8, 0.8, 0.9)	(0.8, 0.9, 1, 1)	(0.7, 0.8, 0.8, 0.9)
	M4	(0.7, 0.8, 0.8, 0.9)	(0.8, 0.9, 1, 1)	(0.7, 0.8, 0.8, 0.9)	(0.7, 0.8, 0.8, 0.9)	(0.7, 0.8, 0.8, 0.9)	(0.8, 0.9, 1, 1)	(0.7, 0.8, 0.8, 0.9)	(0.4, 0.5, 0.5, 0.6)

Table 7.46 Aggregated fuzzy values of material ratings and criterion weightings

	C1	C2	C3	C4	C5	C6	C7	C8
M1	(0.5, 0.64, 0.72, 0.9)	(0.5, 0.76, 0.78, 0.9)	(0.5, 0.64, 0.72, 0.9)	(0.5, 0.68, 0.74, 0.9)	(0.5, 0.64, 0.72, 0.9)	(0.5, 0.72, 0.76, 0.9)	(0.5, 0.64, 0.72, 0.9)	(0.5, 0.64, 0.72, 0.9)
M2	(0.4, 0.62, 0.68, 0.9)	(0.4, 0.58, 0.66, 0.8)	(0.4, 0.7, 0.72, 0.9)	(0.4, 0.66, 0.7, 0.9)	(0.4, 0.7, 0.72, 0.9)	(0.4, 0.62, 0.68, 0.9)	(0.4, 0.58, 0.6, 0.9)	(0.4, 0.6, 0.64, 0.9)
M3	(0.7, 0.86, 0.92, 1)	(0.7, 0.86, 0.92, 1)	(0.7, 0.84, 0.88, 1)	(0.7, 0.84, 0.88, 1)	(0.7, 0.84, 0.88, 1)	(0.7, 0.86, 0.92, 1)	(0.7, 0.84, 0.88, 1)	(0.7, 0.82, 0.84, 1)
M4	(0.7, 0.8, 0.8, 0.9)	(0.7, 0.86, 0.92, 1)	(0.7, 0.84, 0.88, 1)	(0.7, 0.8, 0.8, 0.9)	(0.7, 0.84, 0.88, 1)	(0.7, 0.84, 0.88, 1)	(0.7, 0.86, 0.92, 1)	(0.4, 0.78, 0.82, 1)

Table 7.47 Normalized matrix

	C1	C2	C3	C4	C5	C6	C7	C8
M1	(0.5, 0.64, 0.72, 0.9)	(0.5, 0.76, 0.78, 0.9)	(1.25, 1.60, 1.80, 2.25)	(0.5, 0.68, 0.74, 0.9)	(1.25, 1.60, 1.80, 2.25)	(1.25, 1.80, 1.90, 2.25)	(1.25, 1.60, 1.80, 2.25)	(1.25, 1.60, 1.80, 2.25)
M2	(0.4, 0.62, 0.68, 0.9)	(0.4, 0.58, 0.66, 0.8)	(1.00, 1.75, 1.80, 2.25)	(0.4, 0.66, 0.7, 0.9)	(1.00, 1.75, 1.80, 2.25)	(1.00, 1.55, 1.70, 2.25)	(1.00, 1.45, 1.50, 2.25)	(1.00, 1.50, 1.60, 2.25)
M3	(0.7, 0.86, 0.92, 1)	(0.7, 0.86, 0.92, 1)	(1.75, 2.10, 2.20, 2.50)	(0.7, 0.84, 0.88, 1)	(1.75, 2.10, 2.20, 2.50)	(1.75, 2.15, 2.30, 2.50)	(1.75, 2.10, 2.20, 2.50)	(1.75, 2.05, 2.10, 2.50)
M4	(0.7, 0.8, 0.8, 0.9)	(0.7, 0.86, 0.92, 1)	(1.75, 2.10, 2.20, 2.50)	(0.7, 0.8, 0.8, 0.9)	(1.75, 2.10, 2.20, 2.50)	(1.75, 2.10, 2.20, 2.50)	(1.75, 2.15, 2.30, 2.50)	(1.00, 1.95, 2.05, 2.50)

Table 7.48 **Crisp values for weighting and material ratings**

	C1	C2	C3	C4	C5	C6	C7	C8
W	0.06	0.06	0.15	0.06	0.18	0.15	0.18	0.15
M1	0.69	0.72	1.73	0.70	1.73	1.79	1.73	1.73
M2	0.65	0.61	1.68	0.66	1.68	1.63	1.57	1.60
M3	0.87	0.87	2.13	0.85	2.13	2.16	2.13	2.11
M4	0.8	0.87	2.13	0.80	2.13	2.13	2.16	1.84

A greater focus on fuzzy MADM could produce interesting findings; therefore the method presented in this section is the subject of further consideration in the expectation of a greater degree of accuracy on materials selection using fuzzy data.

7.12   Materials selection for hard surface coatings

One of the main leading causes of failure in many mechanical products or machines is the wear occurring between moving components in contact with each other after frequent or long use. Some wear is usually unavoidable during normal operating conditions, however, excessive wear results in a waste of energy and a reduction of the machine's efficiency, consequently raising the overall cost. Therefore, it is important to avoid or reduce the wear rate to a minimum, which can be done by changing the physical properties of the mating surfaces. Applying hard coating materials is one of the well-known ways of improving the wear resistant properties, for example, on cutting tools. Traditionally, hardness was known as the single most important characteristic for wear resistance. However, the materials for hard coating applications should possess several specific characteristics to maintain the function during operation. The most important properties include hardness (H), Young's modulus (E), adhesion to the substrate, friction coefficient and thermal expansion (α). Furthermore, the H/E and H^3/E^2 ratios are measures of recovery resistance and plastic deformation resistance respectively that influence the wear of the coating material and are also important. The relationship between recovery resistance (Rs) and H/E is given by Eq. (7.23), as follows:

$$Rs = fn\left(\frac{E_r}{H}\right) \tag{7.23}$$

Here, fn signifies function of, and E_r stands for the reduced elastic modulus, which is obtained by:

$$\frac{1}{E_r} = \frac{1 - \nu_1^2}{E_1} + \frac{1 - \nu_2^2}{E_2} \tag{7.24}$$

where ν_1, ν_2, E_1, and E_2 are Poisson's ratios and elastic moduli of indenter and specimen, respectively. A lower resistance to recovery means a lower wear rate and from Eq. (7.24) the ratio of H/E should be higher. The relationship between resistance to plastic deformation or yield pressure, (Py) and H^3/E^2 is indicated by Eq. (7.25) as follows:

$$Py = 0.78 \, r^2 \, \frac{H^3}{E_r^2} \tag{7.25}$$

In the above Equation, r stands for the radius of a contacting sphere. It can be seen from Eq. (7.25), that a higher ratio of H^3/E_r^2 results in higher resistance to plastic deformation and subsequently lower wear rate. It should be noted that E_r can be replaced by E of the material concerned when the loading pair may not be predictable or may vary during the whole life-cycle of the component. Here, in this example, E is used instead of E_r in all the calculations.

For some of the properties, including hardness, Young's modulus, and adhesion, higher values are desirable. Nevertheless, for some other properties such as friction coefficient and thermal expansion coefficient, lower values are preferable. Under operation conditions, where a repeated cyclic load is applied, the temperature of the surface gradually increases, however it is noticeable and after long-time use or continuous running conditions, the components start getting hot. Therefore, the thermal stresses induced can adversely affect the mechanical properties of the component. A lower thermal expansion coefficient can better resist wear and plastic deformation. There is unfortunately no coating material that satisfies all the requirements and each material performs differently for each property. Therefore, it is required to choose an alternative coating material, which can satisfy all the desired properties to the highest degree.

In this example, hardness, elastic modulus, thermal expansion coefficient, H/E and H^3/E^2 are considered as the selection criteria for ranking of 37 candidate coating materials. The TOPSIS method is used to rank the competing coating materials based on the desirable attributes (Chauhan and Vaish, 2013). For the TOPSIS method, the positive ideal solution or most desirable solution is identified by considering the maximum value of hardness, and the minimum values of thermal expansion coefficient and elastic modulus. In contrast, for the negative ideal solution or least desirable solution is determined by considering the minimum value of hardness, and the maximum values of thermal expansion coefficient and elastic modulus. Furthermore, for the two other criteria (H/E, H^3/E^2), higher values are also desirable. Identical weightings for all the attributes are taken into account. Table 7.49 shows the five criteria used, candidate materials and the ranking order obtained by the TOPSIS method.

The results show that carbon (C) is the most desirable candidate coating material in the ranking scheme, while MgO is the least desirable candidate coating material. Also, the ranking order indicates that covalent bonded coating materials have in general the highest rankings, which possibly explains why these materials perform competitively better than other materials for hard coating applications.

Table 7.49 Hard coating materials—properties and TOPSIS ranking

Materials	Hardness (HV)	Elastic modulus (GPa)	Thermal expansion coefficient (α) (10^{-6} K^{-1})	H/E	H^3/E^2	TOPSIS index	Ranking
C	8000	910	1	8.79121	618,282.8161	0.999999999	1
BN	5000	660	1.2	7.57576	286,960.5142	0.539721414	2
B_4C	3500	441	4.5	7.93651	220,458.5538	0.41471789	3
Si_3N_4	1720	210	2.5	8.19048	115,384.3084	0.305588339	4
VC	2900	430	7.3	6.74419	131,903.7318	0.284850999	5
SiB_6	2300	330	5.4	6.9697	111,726.3545	0.271657875	6
LaB_6	2530	400	6.4	6.325	101,214.2313	0.252038492	7
ZrC	2560	400	7.2	6.4	104,857.6	0.249325945	8
SiC	2600	480	5.3	5.41667	76,284.72222	0.237641253	9
TiC	2800	470	8.3	5.95745	99,375.28293	0.23756736	10
TiB_2	3000	560	7.8	5.35714	86,096.93878	0.231603751	11
B	2700	490	8.3	5.5102	81,978.34236	0.215416952	12
WC	2350	720	3.85	3.26389	25,034.4811	0.207270902	13
CrN	1100	400	2.3	2.75	8318.75	0.200703518	14
ZrB_2	2300	540	5.9	4.25926	41,724.96571	0.190860318	15
NbB_2	2600	630	8	4.12698	44,283.19476	0.175515216	16
Al_2O_3	2100	400	8.4	5.25	57,881.25	0.174923476	17
W_2B_5	2700	770	7.8	3.50649	33,197.84112	0.17170734	18
VB_2	2150	510	7.6	4.21569	38,209.82314	0.161757395	19
Cr_3C_2	2150	400	11.7	5.375	62,114.84375	0.156424059	20
AlN	1230	350	5.7	3.51429	15,190.75102	0.153565153	21
ZrO_2	1200	190	11	6.31579	47,867.03601	0.146876763	22
Mo_2B_5	2350	670	8.6	3.50746	28,910.39207	0.14602969	23
NbC	1800	580	7.2	3.10345	17,336.50416	0.192729	24

TaB$_2$	2100	680	8.2	3.08824	20,028.11419	0.135182479	25
HfO$_2$	780	220	6.5	3.54545	9804.79339	0.134920417	26
CrB$_2$	2250	540	10.5	4.16667	39062.5	0.133925289	27
ZrN	1600	510	7.2	3.13725	15,747.78931	0.133633183	28
TiO$_2$	1100	205	9	5.36585	31,671.62403	0.133571595	29
TaC	1550	560	7.1	2.76786	11,874.6014	0.131361551	30
TiN	2100	590	9.4	3.55932	26,604.42402	0.12553725	31
Mo$_2$C	1660	540	8.05	3.07407	15,686.88615	0.121103826	32
BeO	1500	390	9	3.84615	22,189.34911	0.111201036	33
VN	1560	460	9.2	3.3913	17,941.47448	0.103172246	34
ThO$_2$	950	240	9.3	3.95833	14,884.98264	0.095149553	35
NbN	1400	480	10.1	2.91667	11,909.72222	0.077772706	36
MgO	750	320	13	2.34375	4119.87305	$4.27912E - 10$	37

7.13   Summary and conclusions

A variety of practical engineering design problems demonstrates the scope
and application of different decision-making techniques and quality tools available
for materials selection and design. The problem-solving process is complex and
places significant demands on the part of designers to strategically employ the
most appropriate tools and techniques. As well as the large amount of conflicting
materials and design information needed to be considered, uncertainty in data is
one of the main issues that inhibits decision-making. In materials selection, the use
of unreliable data for cost estimating is also unavoidable and this necessitates
suitable decision-making methods for design selection. Further effort therefore
needs to be made to consider material design and selection, and design optimization
simultaneously in the engineering design process.

Review questions

1. What are the possible ways of obtaining objective and dependency weightings for interval
   data?
2. In the materials selection for a knee prosthesis example:
   a. Calculate the objective weightings.
   b. Obtain the weighting of criteria using the revised Simos method.
   c. Combine the objective and subjective weightings, with the assumption of uncertainty
      in importance of each type of weightings.
   d. Obtain the ranking order of materials for different weightings and interpret the final
      ranking of materials using a graph that shows frequency of each material to each rank
      (eg, if you obtain the ranking of materials ten times and Material 1 assigns to rank 8, 7
      times; then you might interpret Material 1 has rank 8 with 70% confidence).
3. In the materials selection for an aircraft patch repair example:
   a. Obtain the weighting of dependency and objective.
   b. Determine the design target values and importance of criteria using QFD.
   c. Combine the three types of weightings with the assumption of equal importance on
      each type of weightings and calculate the final ranking of materials.
   d. Using the middle value of interval data and weighting obtained in the last step;
      determine the ranking orders of materials by both comprehensive VIKOR method and
      target-based TOPSIS method.
   e. If it is applicable, obtain the optimum ranking of materials by the aggregation method.
   f. If a designer decides to consider only composite materials for the repair, do you think
      you should recalculate the ranking of materials and why?
4. In the materials selection for a turnbuckle example:
   a. Discuss the advantages and disadvantages of using Pareto points.
   b. Discuss the reasons for cost uncertainties in the materials selection process.
5. In the materials and design selection for thin-walled lightweight cylindrical tubes:
   a. Evaluate the sensitivity of ranking orders to weighting of criteria.
   b. Rank design alternatives base on average value of experiments and simulation, and
      then compare the results with interval data method.

6. In the material selection for hard coating example:

  a. Rank the materials using material properties rather than performance indices, and compare the results.

  b. Discuss the accuracy of the aforementioned approaches.

References

Asiedu, Y., Gu, P., 1998. Product life cycle cost analysis: state of the art review. Int. J. Prod. Res. 36, 883–908.

Bahraminasab, M., Jahan, A., 2011. Material selection for femoral component of total knee replacement using comprehensive VIKOR. Mater. Des. 32, 4471–4477.

Bahraminasab, M., Sahari, B.B., Edwards, K.L., Farahmand, F., Hong, T.S., Naghibi, H., 2013. Material tailoring of the femoral component in a total knee replacement to reduce the problem of aseptic loosening. Mater. Des. 52, 441–451.

Bahraminasab, M., Sahari, B., Edwards, K.L., Farahmand, F., Hong, T.S., Arumugam, M., et al., 2014a. Multi-objective design optimization of functionally graded material for the femoral component of a total knee replacement. Mater. Des. 53, 159–173.

Bahraminasab, M., Sahari, B.B., Edwards, K.L., Farahmand, F., Jahan, A., Hong, T.S., et al., 2014b. On the influence of shape and material used for the femoral component pegs in knee prostheses for reducing the problem of aseptic loosening. Mater. Des. 416–428.

Baker, A.A., Rose, L.R.F., Jones, R., 2002. Advances in the Bonded Composite Repair of Metallic Aircraft Structure. Oxford, Elsevier.

Chauhan, A., Vaish, R., 2013. Hard coating material selection using multi-criteria decision making. Mater. Des. 44, 240–245.

Chen, S.J.J., Hwang, C.L., Beckmann, M.J., Krelle, W., 1992. Fuzzy Multiple Attribute Decision Making: Methods and Applications. Springer-Verlag New York, Inc., Secaucus, NJ.

Dehghan-Manshadi, B., Mahmudi, H., Abedian, A., Mahmudi, R., 2007. A novel method for materials selection in mechanical design: combination of non-linear normalization and a modified digital logic method. Mater. Des. 28, 8–15.

Del Castillo, E., 1996. Multiresponse process optimization via constrained confidence regions. J. Qual. Technol. 28, 61–70.

Del Castillo, E., Montgomery, D.C., 1993. A nonlinear programming solution to the dual response problem. J. Qual. Technol. 25, 199–204.

Derringer, G., Suich, R., 1980. Simultaneous optimization of several response variables. J. Qual. Technol. 12, 214–219.

Duong, C.N., Wang, C.H., 2007. Composite Repair: Theory and Design. Oxford, Elsevier.

Farag, M.M., 1997. Materials Selection for Engineering Design. Prentice-Hall, New York, NY.

Farag, M.M., 2002. Quantitative methods of materials selection. In: Kutz, M. (Ed.), Handbook of Materials Selection. London, John Wiley & Sons.

Farag, M.M., 2008. Materials and Process Selection for Engineering Design. CRC Press Taylor and Francis Group, London.

Fayazbakhsh, K., Abedian, A., 2009. Materials selection for applications in space environment considering outgassing phenomenon. Adv. Space Res. 45, 741–749.

Fouladi, E., Fayazbakhsh, K., Abedian, A., 2010. Patch materials selection for ageing metallic aircraft structures using digital quantitative materials selection methods. In: 27th International Congress of the Aeronautical Sciences, Nice, France.

Gibson, L.J., Ashby, M.F., 1999. Cellular Solids: Structure and Properties. Cambridge, Cambridge Univ. Press.

Goh, Y.M., Newnes, L.B., Mileham, A.R., Mcmahon, C.A., Saravi, M.E., 2010. Uncertainty in through-life costing-review and perspectives. Eng. Manage., IEEE Trans. 57, 689−701.

Jahan, A., Bahraminasab, M., 2015. Multicriteria decision analysis in improving quality of design in femoral component of knee prostheses: influence of interface geometry and material. Adv. Mater. Sci. Eng. 2015, 16.

Jahan, A., Edwards, K.L., 2013. VIKOR method for material selection problems with interval numbers and target-based criteria, 47, 759−765.

Jahan, A., Ismail, M.Y., Shuib, S., Norfazidah, D., Edwards, K.L., 2011a. An aggregation technique for optimal decision-making in materials selection. Mater. Des. 32, 4918−4924.

Jahan, A., Mustapha, F., Ismail, M.Y., Sapuan, S.M., Bahraminasab, M., 2011b. A comprehensive VIKOR method for material selection. Mater. Des. 32, 1215−1221.

Jahan, A., Bahraminasab, M., Edwards, K.L., 2012. A target-based normalization technique for materials selection. Mater. Des. 35, 647−654.

Jee, D.H., Kang, K.J., 2000. A method for optimal material selection aided with decision making theory. Mater. Des. 21, 199−206.

Jeya Girubha, R., Vinodh, S., 2012. Application of fuzzy VIKOR and environmental impact analysis for material selection of an automotive component. Mater. Des. 37, 478−486.

Karande, P., Gauri, S.K., Chakraborty, S., 2012. Applications of utility concept and desirability function for materials selection. Mater. Des. 45, 349−358.

Kurtz, S., Ong, K., Lau, E., Mowat, F., Halpern, M., 2007. Projections of primary and revision hip and knee arthroplasty in the United States from 2005 to 2030. J. Bone Joint Surg. - Ser. A. 89, 780−785.

Kutz, M., 2002. Handbook of Materials Selection. London, John Wiley & Sons.

Leite, M., Silva, A., Henriques, E., Madeira, J., 2015. Materials selection for a set of multiple parts considering manufacturing costs and weight reduction with structural isoperformance using direct multisearch optimization. Struct. Multidiscip. Optim. 1−10.

Lemaire, R., 2010. Fatigue fracture of the femoral component in a mobile bearing knee prosthesis. Acta. Orthop. Belg. 76, 274−281.

Levine, B., Fabi, D., 2010. Porous metals in orthopedic applications—a review. Materwiss. Werksttech. 41, 1001−1010.

Marsh, G., 2014. Patching them up, the composites way. Reinf. Plast. 58, 40−44.

Montgomery, D.C., Runger, G.C., 2010. Applied Statistics and Probability for Engineers. John Wiley & Sons.

Noorossana, R., Ardakani, M.K., 2009. A weighted metric method to optimize multi-response robust problems. J. Ind. Eng. Int. 5, 10−19.

Oonishi, H., Kim, S., Kyomoto, M., Iwamoto, M., Ueno, M., 2006. PE wear in ceramic/PE bearing surface in total knee arthroplasty: Clinical experiences of more than 24 years. Bioceramics Alternative Bearings in Joint Arthroplasty. pp. 101−110, Steinkopff.

Park, J.B., Bronzino, J.D., 2003. Biomaterials: Principles and Applications. CRC, Boca Roton, FL.

Pasandideh, S.H.R., Niaki, S.T.A., 2006. Multi-response simulation optimization using genetic algorithm within desirability function framework. Appl. Math. Comput. 175, 366−382.

Ramakrishna, S., Mayer, J., Wintermantel, E., Leong, K.W., 2001. Biomedical applications of polymer-composite materials: a review. Composites Sci. Technol. 61, 1189−1224.

Rao, R.V., 2006. A material selection model using graph theory and matrix approach. Mater. Sci. Eng. A. 431, 248−255.

Rao, R.V., Davim, J.P., 2008. A decision-making framework model for material selection using a combined multiple attribute decision-making method. Int. J. Adv. Manuf. Technol. 35, 751−760.

Rezvani, M., Jahan, A., 2015. Effect of initiator, design, and material on crashworthiness performance of thin-walled cylindrical tubes: a primary multi-criteria analysis in lightweight design. Thin Wall Struct. 96, 169−182.

Shanian, A., Savadogo, O., 2006. A non-compensatory compromised solution for material selection of bipolar plates for polymer electrolyte membrane fuel cell (PEMFC) using ELECTRE IV. Electrochim. Acta. 51, 5307−5315.

Shanian, A., Milani, A.S., Carson, C., Abeyaratne, R.C., 2008. A new application of ELECTRE III and revised Simos' procedure for group material selection under weighting uncertainty. Knowl. Based Syst. 21, 709−720.

Teoh, S.H., 2000. Fatigue of biomaterials: a review. Int. J. Fatigue. 22, 825−837.

Thurston, D.L., Locascio, A., 1994. Decision theory for design economics. Eng. Econ. 40, 41−71.

Wu, G., Yang, J.M., 2005. The mechanical behavior of GLARE laminates for aircraft structures. JOM J. Miner., Met. Mater. Soc. 57, 72−79.

Future developments

8

Learning Aims

The overall aim of this chapter is to discuss a number of emerging and potential future research areas related to Multi-Criteria Decision-Making (MCDM) in materials and design. After carefully studying this chapter you should be able to understand:

- The current position of MCDM in design and materials selection.
- The future directions for developing MCDM techniques for solving design problems.
- The possible applications of MCDM in materials design and selection.

8.1 Overview of the current situation

Materials selection is an integral part of the design process and to be fully effective it has to also take into consideration the interdependency of associated materials and manufacturing processes. However, it is difficult for designers, even when educated in the fundamentals of materials and mechanical engineering, to still be able to make optimum decisions on the choice of materials to satisfy design problems given the vast range of available materials and new materials being developed. Today, new materials are being developed faster than at any other time in history, therefore the challenges and opportunities are greater than ever before (Ashby, 2005).

The previous chapters have demonstrated the various methods that are either currently being used or are in the process of being developed for supporting designers in choosing the most suitable materials for new products. MCDM has the potential to improve all areas of decision-making in engineering, from design to manufacture but is especially useful for high technology market sectors such as aerospace, electronics, motorsports, nuclear, and biomedical applications, where product differentiation and competitive advantage are often achieved by just very small gains in material performance.

The following sections emphasize the areas of future development that are desirable for improving materials selection in the design of engineering products, highlighting aspects that are either already in progress or intended by the authors for strategically applying MCDM methods.

Multi-criteria Decision Analysis. DOI: http://dx.doi.org/10.1016/B978-0-08-100536-1.00008-4

8.2   The development of decision-making methods for actual design scenarios

Researchers have developed and promoted a broad range of decision-making tools and techniques that have been widely taught and that are in widespread use by designers for supporting design selection. However, a lot of these tools are not consistent with the rigorous principles of decision theory, and these inconsistencies lead to undesirable behavior and therefore outcomes of engineering design methods. Often, such undesirable behavior is hidden by the complexity of the design process, and can lead to poor design decision-making in a way that goes unrecognized by designers (Hazelrigg, 2003). It is therefore desirable to develop improved systematic decision-making methods that fully take into consideration "real" engineering design problems that involve conflict and compromise in the engineering design process and uncertainty in the materials and design information/data available.

- A large number of methods, including materials selection methods, have been developed for guiding the designer to an optimum decision. The decision support methods have improved the quality of decision-making but as expected there are advantages and disadvantages with each method. The limitations of the methods are often as a consequence of their development and use in isolation of other methods and with using incomplete information. There is therefore a need for better integration of decision models with a more complete characterization of engineering design problems. It is recommended that embodying the outcomes from real/practical design studies will help speed up the progress.
- The best design solution can only be determined if selection decisions for materials, shape and processing are all considered simultaneously. As a result, future research should therefore not only concentrate on supporting the multiple consideration activity but also on investigating the effects of the inevitable need for compromise in decision-making that needs to be made between these different areas of materials selection by using Multiple Attribute Decision-Making (MADM).
- Identifying innovative products by exploiting newly developed materials is essential for gaining competitive and time-to-market benefits. As a consequence, the application of MADM for discovering promising applications of new materials is an important topic for future research (Sirisalee et al., 2004).
- One of the most important reasons that materials selection must be based on interval data, as opposed to discrete data, is that material properties differ so widely in practice, that is, subject to tolerances, as a consequence of manufacturing process variability. Uncertainty is not only emerging as a research field in materials selection but also in materials design applications. Stochastic uncertainty stems from stochastic variability and inherent randomness of material processing and morphology. McDowell and Olson (2009), for example, highlight the role of uncertainty in materials design and cite the need for robust materials design solutions that are relatively insensitive to variations in material structure at various scales.
- In the area of materials selection, the interval numbers neither indicate how probable it is to the value to be in the interval, nor specify which of the many values in the interval is the most likely to occur. Therefore, considering the probabilistic distribution function of materials properties is important in the materials selection process.

- Materials engineering decisions have economic consequences with regards to production costs and profitability. The designer might strive to design the perfect component, but the price must be competitive in the commercial marketplace. Therefore, in engineering practice in addition to scientific principles, economic criteria must be considered in the development of a marketable product. However, the uncertainty associated with cost estimating, is one of the main limitations of the materials selection process.
- The need for a diversity of decision-making methods for tackling different design problems has encouraged researchers to develop new techniques. Selecting optimal materials, shapes, and processes for the design of components from a range of proposed alternatives via MCDM helps designers and, therefore, manufacturers to develop high-quality products at acceptable cost.
- The need to consider a large number of possible materials adds to the difficulties in making a selection. In practice, this often leads to compromise in terms of the number of materials considered (limited to a number of familiar materials only) and causes some of the design requirements not to be entirely satisfied (partially met only). MCDM allows many materials and design attributes to be considered simultaneously, leading to more successful identification of suitable materials that meet the desired design requirements. However, there are fewer tools available that are dedicated to materials selection, although many design tools make use of materials properties databases.
- For the future, the ranking procedure can be implemented in computer code and interfaced with commercially available materials databases software for more rapid analysis of different design problems and material combinations.

8.3 The application of MCDM methods to complex materials selection and design problems

There is a need to be able to apply MCDM methods to more complex materials selection problems involving functionally graded materials, multi-phase materials, and multiple materials based components.

- Selecting an optimal refinement condition to achieve the best performance from composite materials (Khorshidi et al., 2013) is an example of materials selection that is in need of further research. The judicious choice of constituent materials (reinforcement and matrix) and their proportions (volume fraction) and geometry/orientation (continuous, discontinuous, particulate, etc.) that may need to vary throughout the structure of a component, is nontrivial and difficult to achieve without proper decision support.
- The selection of materials is a decision made in the early-stage of the new product development process, but has a significant influence on the overall life cycle of the final product. Using a traditional engineering economic analysis it is possible to convert the required capital investment and cash flow for each alternative material to a common metric using equivalent annual cost or net present value analyses at the appropriate interest rate. However, the "time value of money" alone fails to fully reflect the way in which manufacturing concerns view the trade-off between capital investment requirements and variable costs. This is of high importance for industries in which investment in capital equipment needs significant long-term commitment. For this reason, it is sometimes necessary to define "capital cost" and "variable cost" as separate and distinct attributes.

Here, traditional engineering economic analysis for materials selection can be replaced by MADM to consider different technical and economic criteria simultaneously.

- To make something out of a material you also need a material process to be compatible with the material you plan to use. In addition, it is also expected that the matching process helps to dominate competitors' products for performance, economy, and efficiency, and to avoid any damage to the environment (Ashby et al., 2013). It appears that design will benefit greatly from the appropriate incorporation of decision sciences into making the correct choice for the triad relationship between material, processing and design.

- Integrating a materials' property database with design algorithms and computer-aided design/manufacturing (CAD/CAM) programs has a lot of advantages including homogenization and sharing of data, decreased redundancy of effort, and decreased cost of information storage and retrieval (Farag, 2002). Rapid advances in Finite Element Analysis (FEA) methods have also allowed the application of computer simulation as part of an optimization strategy for the design of composite structures. These approaches are useful when a single design objective can be defined but are less useful when there are conflicts in design objectives, for example, minimum weight and cost (Aceves et al., 2008).

- There is a lot of emphasis being placed on the importance of modeling in materials design. It might be feasible in future versions of computer simulation software for MCDM and simulation tools to be combined and take into consideration the role of materials selection in the concurrent design of materials and products. This will allow designers to be able to readily evaluate various materials with different properties, performance indices and costs simultaneously.

- Another possible area of future research would be materials selection for multiple components as opposed to single components. In such situations, usually the optimal solution for the whole product, assembly or group of components is not only complicated to resolve but cannot be easily reached. Therefore, as the number of components increases, along with the number of alternative materials for each component, an automated MCDM process for materials selection is considered to be necessary (Leite et al., 2015).

8.4   Final remarks on trends and challenges for the future

There are many evolving areas of development and applications of MCDM involving materials selection as a consequence of growing consumer demand for higher performance products at reasonable cost.

- It is recognized that people are consuming materials more rapidly than ever but they are also using an increasing diversity of materials. Consequently, engineers and designers are always on the lookout for new materials and improved processes to manufacture better products more efficiently. The selection of the most appropriate materials not only affects the capability of manufacturing systems and satisfaction of customers but also impacts on environmental issues, including recycling.

- The evaluation of a large number of design requirements (size, shape, mass, surface finish, cost, etc.) for each component in a typical product and the suitability of an even larger number of different materials rapidly becomes too complicated to be intuitive, even for the most experienced designers. This highlights the value of being able to use MCDM to support decision-making in the engineering design process. Although, experimental-based

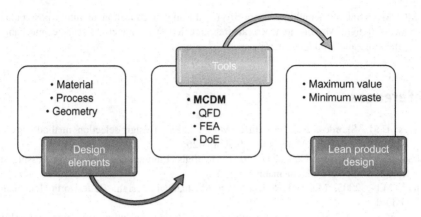

Figure 8.1 MCDM used as a tool for Lean product design.

selection of a material, for example, testing, prototyping, etc., for a specific design solution is commonly used, it quickly becomes unreasonable due to the time required and the high costs of experiments, especially if several different materials have to be considered. Alternatively, other more viable options can be considered such as computer based simulations, but the ranking of materials should have already been successfully completed during the initial stage of the design process. As a consequence, materials selection methods based on MCDM, especially when used for sustainable product design, are growing dramatically in theory and practice. Experimental methods may still be required for validation purposes but their wide scale regular use will be unnecessary as confidence grows in the capability of MCDM methods.

- In the product design process, MCDM can help decrease risk. It is also as a means to increase value will improve the likelihood of delivering the required product specification to the required schedule as shown in Fig. 8.1. In this process, one way to limit unnecessary costs or low performance is to use a well formulated materials selection method. Systematic approaches that enable designers to determine and map the correlation between the different functional specifications and design parameters, is highly demanded. It seems that applying a combined Quality Function Deployment—Finite Element Analysis—Multi-criteria Decision-making—Design of Experiment (QFD—FEA—MCDM—DoE) approach enables appropriate materials to be matched more accurately to design requirements. As a consequence, it is anticipated to:

1. Reduce development costs of new products.
2. Increase customer satisfaction, revenue, and profit.
3. Shorten development times and time-to-market.
4. Lower manufacturing costs.
5. Lessen early-life failures.

However, the traditional design and decision-making process must first be improved before better products can be fully realized.

- In view of the role of materials selection as a bridge between the three fields of knowledge: materials science, mechanical engineering, and product design (Silva, 2015), it is becoming extremely difficult to ignore the importance of related decision-making tools and techniques. Therefore, ensuring appropriate teaching (theory and application) of

MCDM techniques should be a priority for the higher education of future postgraduate students (taught Masters to Doctorate research level) on materials science, mechanical engineering, and product design courses.

References

Aceves, C.M., Skordos, A.A., Sutcliffe, M.P.F., 2008. Design selection methodology for composite structures. Mater. Des. 29, 418−426.

Ashby, M., Shercliff, H., Cebon, D., 2013. Materials: Engineering, Science, Processing and Design. Butterworth-Heinemann, Oxford.

Ashby, M.F., 2005. Materials Selection in Mechanical Design. Butterworth-Heinemann, Oxford.

Farag, M.M., 2002. Quantitative methods of materials selection. In: Kutz, M. (Ed.), Handbook of Materials Selection. Wiley, New York, pp. 3−26.

Hazelrigg, G.A., 2003. Validation of engineering design alternative selection methods. Eng. Optim. 35, 103−120.

Khorshidi, R., Hassani, A., Honarbakhsh Rauof, A., Emamy, M., 2013. Selection of an optimal refinement condition to achieve maximum tensile properties of Al-15%Mg$_2$Si composite based on TOPSIS method. Mater. Des. 46, 442−450.

Leite, M., Silva, A., Henriques, E., Madeira, J., 2015. Materials selection for a set of multiple parts considering manufacturing costs and weight reduction with structural isoperformance using direct multisearch optimization. Struct. Multidisciplinary Optim. 52, 635−644.

McDowell, D.L., Olson, G.B., 2009. Concurrent design of hierarchical materials and structures. Sci. Model. Simulations. 15, 207−240.

Silva, A., 2015. Materials as a bridge between science, engineering, and design. In: Lim, H.W. (Ed.), Handbook of Research on Recent Developments in Materials Science and Corrosion Engineering Education. IGI Global, Hershey, PA, pp. 292−308.

Sirisalee, P., Ashby, M.F., Parks, G.T., Clarkson, P.J., 2004. Multi-criteria material selection in engineering design. Adv. Eng. Mater. 6, 84−92.

Index

Printed in the United States
By Bookmasters